"In this highly readable and fascinating book, Dru Johnson offers readers an intriguing, imaginative, and eye-opening account that opens a window into two disparate worlds—a world of evolutionary science and a world of (some of) the biblical authors—while also showing them to be, in many ways, complementary to one another. What is especially laudable is Johnson's balanced approach that succeeds in being at once both cautious and creative."
Andrew Torrance, University of St. Andrews

"Johnson raises distinctly relevant considerations about the historical and future nature of reality and the distinctions leaving biblical and evolutionary narratives in tension. I will be pondering much of what Johnson has to say for years to come and look forward to conversations his contributions invite others to. I revel in the idea of understanding some things anew and pondering deeper things still in tension, and I appreciate Johnson's faithful scriptural exegesis to aid me in my considerations. Can evolutionary and biblical narratives converge, as Christians who embrace evolution's explanations of origins claim? Or are persistent, irreconcilable tensions inevitable as one metaphysical position clings to an embedded and inescapable narrative of exclusion, competition, scarcity, and existential insecurity?"
Anjeanette "AJ" Roberts, molecular biologist, author, and chaplain

"What hath Eden to do with the Galápagos Islands? As Dru Johnson explains, more than one might expect. We who live in a culture of affluence have difficulty grasping how powerful the message of Edenic flourishing would have been to the original audience, a culture threatened constantly by scarcity and violence. Johnson demonstrates that Moses and Darwin dealt with many of the same subjects but came to very different conclusions."
Kenneth Keathley, senior professor of theology at Southeastern Baptist Theological Seminary

"In this book, Dru Johnson offers readers a brilliant thought experiment: suppose we compare Darwin and the authors of Scripture on the intersection of scarcity, sex, and environmental fit—topics that were clearly important for Darwin's understanding of the evolutionary process. It turns out that these topics are central also to the way Scripture portrays Eden, life after the fall, and the new heavens and new earth. Although this book does not (by a long shot) 'harmonize' Scripture and science, the thematic comparison generates many exegetical insights and sheds significant light on the Bible's vision of God's intent for creation."
J. Richard Middleton, professor of biblical worldview and exegesis at Northeastern Seminary, Roberts Wesleyan University

DRU JOHNSON

WHAT HATH DARWIN TO DO WITH SCRIPTURE?

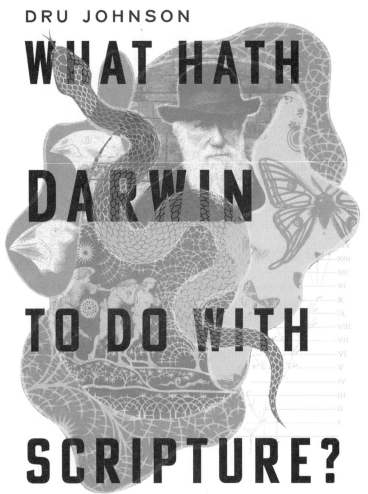

COMPARING THE CONCEPTUAL WORLDS
OF THE BIBLE AND EVOLUTION

ivp

Academic

An imprint of InterVarsity Press
Downers Grove, Illinois

InterVarsity Press
P.O. Box 1400 | Downers Grove, IL 60515-1426
ivpress.com | email@ivpress.com

InterVarsity Press® is the publishing division of InterVarsity Christian Fellowship/USA®. For more information, visit intervarsity.org.

Scripture quotations, unless otherwise noted, are from The Holy Bible, English Standard Version, copyright © 2001 by Crossway Bibles, a division of Good News Publishers. Used by permission. All rights reserved.

Figure 10.1: From Charles Darwin, *On the Origin of Species*, 1859. Digital picture taken by Alexei Kouprianov / Wikimedia Commons.

The publisher cannot verify the accuracy or functionality of website URLs used in this book beyond the date of publication.

Cover design: David Fassett
Interior design: Daniel van Loon
Images: Getty Images: © William Harbison / EyeEm, © mikroman6, © Sergey Ryumin, © Photos.com, © Marina Tikhoplav, © vinap; Wikimedia Commons: The Expulsion from Eden / National Gallery of Art, Washington / Raw Pixel, Charles Darwin in 1881

ISBN 978-1-5140-0361-9 (print) | ISBN 978-1-5140-0362-6 (digital)

Printed in the United States of America ∞

Library of Congress Cataloging-in-Publication Data

Names: Johnson, Dru, author.
Title: What hath Darwin to do with scripture? : comparing conceptual worlds
   of the Bible and evolution / Dru Johnson.
Description: Downers Grove, IL : IVP Academic, [2023] | Includes
   bibliographical references and index.
Identifiers: LCCN 2023025368 (print) | LCCN 2023025369 (ebook) | ISBN
   9781514003619 (print) | ISBN 9781514003626 (digital)
Subjects: LCSH: Bible and evolution. | Evolution–Religious
   aspects–Christianity. | Creationism.
Classification: LCC BS659 .J573 2023 (print) | LCC BS659 (ebook) | DDC
   231.7/652–dc23/eng/20230719
LC record available at https://lccn.loc.gov/2023025368
LC ebook record available at https://lccn.loc.gov/2023025369

CONTENTS

ACKNOWLEDGMENTS

THIS BOOK WAS RESEARCHED and written mostly while I was a senior research fellow at the Carl Henry Center of Trinity Evangelical Divinity School in Deerfield, Illinois, under The Creation Project, a Templeton Religion Trust funded grant. I would first like to thank the Henry Center staff, Joel Chopp and Matthew Wiley, for vigorous conversations and insights during my time there. A special note of thanks goes to Geoffrey Fulkerson, the then-director of the Henry Center, without whom this book would not have come into existence. He prodded me to consider researching in this direction and helped me to shepherd my own thoughts along the way. (Even if readers don't believe that I've thought it through enough!) I owe him a deep debt of gratitude.

I also learned so much from the other fellows: Jim Hoffmeier, Ryan Peterson, and Mary Vanden Berg; along with our reading group: Tom McCall, Bradley Gundlach, and Dick Averbeck. Additionally, the Dabar conferences were some of the most intellectually stimulating conferences in which I have participated.

Friends and colleagues critiqued drafts and provided invaluable corrections. A heaping thanks to Joshua Swamidass for his care and corrections. Michael Rhodes, Taylor Lindsay, and Celina Durgin all read and helped to edit and clarify earlier drafts. IVP's anonymous reviewer provided helpful guidance that led to me cutting over twelve thousand words from the manuscript. The reader can thank that reviewer in their hearts.

Jon Boyd, who understood this weird project like no other, has been an amazing editor. He supported me with critique and encouragement the whole way to completion. Jim Kinney, of Baker Academic, was gracious enough to talk through the book idea with me a few times.

Going back twenty-four years, it was Esther Meek, Mike D. Williams, and Jack Collins who all helped me to see the deep convergences between

Scripture and scientific thought when I was a student at Covenant Theological Seminary.

Finally, to Stephanie and our children, Benjamin, Claudia, Olivia, and Luisa, thank you for letting me commute to Chicago every week for fourteen weeks to work on this project. You all gave up more than anyone else for this book, even though you'll never see a dime from the royalties!

PART ONE

TWO CONCEPTUAL WORLDS

CHAPTER ONE

THIS IS NOT THE CREATION-EVOLUTION DEBATE YOU'RE LOOKING FOR

THE CULTURE WARS OF EVOLUTION have set our minds on two paths—more like two ruts. Our creation theologies can ride or die with the Jesus fish or the Darwin fish on their bumpers. Ironically, a devout Christian driving that Jesus-fish car, a vehicle that resulted from centuries of applied science, pooh-poohs "those scientists" for promoting an anti-Christian agenda. Elsewhere in the world, a theoretical physicist naively dismisses religion as blind faith in invisible spirits. I do not want to contribute to the supposed conflict between science and faith. I want to do something much more disruptive. I want us to read Scripture for its own views on natural selection.

The Hebrew creation accounts (specifically Gen 1–2, among others) sew together the same three concerns that Darwin eventually identified as the central topics of natural selection: scarcity of resources, fittedness to habitat, and their combined impacts on sexual propagation.[1] My goal here is to consider the parallels among Darwin's natural selection and later conceptual developments in evolutionary science, then compare them to the conceptual

---

[1]I am fully aware that some might want to work out what is *the single correct view* of creation: what *God actually saw* or *what actually happened* in creation. Others might want to figure out what the biblical authors meant relative to the other creation accounts in the ancient Near East or by analysis of the literary forms and words being used in Genesis. Though good and worthy things to think through, neither of these approaches will get us where we need to go here.

world of the Bible.[2] We will see both where the two views jive and where they diverge.

Even if it is a grand coincidence, the overlapping concern with genealogical selection in Scripture and in Darwin's thinking deserves some attention. After all, both views supply stories about the beginning, middle, and future of the cosmos.

These biblical and scientific folktales mean to speak realistically about our beginnings. Both stories intend to say something true about the natural history of the universe (even if one thinks the biblical authors do so poetically or analogically). Hence, I use the terms *folktale* and *mythology* as positive terms, not pejorative ones. *Folktale* does not relate to scientific or historical value but to explanatory purpose.[3] Like all folktales, these two views about creation are designed to explain what we see and how it came to be.

The biblical authors constructed polished and concise stories of human origins to reason with their audiences (which later includes us) about the invisible and organizing features of our cosmos. Because the biblical literature consistently develops these metaphysical views, we will do well to trace their metaphysical assumptions from beginning to end. The metaphysics that the biblical authors want us to understand is not a complex abstract system but a version of our material world reimagined with a different orientation.

Whether you read biblical texts as history or mystery, they are not just telling stories; they are selling an intellectual tradition rooted in creation. We will soon see the same with some folktales from the evolutionary sciences.

Some of these folktales from the evolutionary sciences differ from the biblical ones on this front: they offer no vision for the way things *ought* to be. For many, if not most, there simply is no particular way in which the

---

[2]By *Bible*, *Scripture*, or *Scriptures*, I am referring to the minimal collection of texts regarded by all Christian traditions as the Bible: both the Hebrew Bible/Old Testament and New Testament.
[3]Many have adequately argued that Genesis does not actually belong in the genre of myth if for no other reason than its rhetorical aims. So James K. Hoffmeier summarizes the comments of Nahum Sarna, Gordan Wenham, and Umberto Cassuto: "The Torah displays an aversion for myth, and as suggested above, combating the ancient Near Eastern mythologies is overtly and subtly at work in the book of Genesis." Hoffmeier, "Genesis 1–11 as History and Theology," in *Genesis: History, Fiction, or Neither: Three Views on the Bible's Earliest Chapters*, ed. Stanley N. Gundry (Grand Rapids, MI: Zondervan, 2015), 23-58, here 40.

material universe is supposed to be oriented. For most versions of the story in the evolutionary sciences, the cosmos *now is* as it always *has been* and ever more *shall be*. Not so for the biblical authors, from the Hebrew Scriptures (Old Testament) to the New Testament. Understanding that pivotal reorientation of the cosmos (what I am calling the metaphysical aspect) illuminates how the biblical authors singularly frame the intersection of scarcity, fit, and sex in the process of genetic selection.

Overlaying the maps of these two intellectual worlds—the Bible's and evolutionary science's—will show us the various routes they each forge to conceptualize the world we know today. The biblical authors' persistent dealings with communitarian ethics, scientific paths to knowledge, metaphysical principles, and causal physical relationships make the biblical intellectual world remarkably relevant for us. Even more, it is relevant for us in ways that other ancient intellectual worlds are not.

I will argue that the intellectual world of the biblical authors makes our world existentially, ethically, and physically coherent in a way that could be harmonized with many of the findings of science—depending on how one construes both enterprises.

INTERRUPTING SCRIPTURE WITH MY QUESTION

"What about the dinosaurs?" I loudly interrupted. The living room was littered with teens on couches around the man who had been speaking to us. "What about them?" the young married youth leader countered. I was twelve, suffering from the angst-ridden effects of my parents' divorce. John Ragsdale happened to be an unfortunate victim of my antics. I was not going to let this youth pastor say another word until he explained to me where the dinosaurs are in the Bible.

Of course, bringing our specific aches and questions to the Bible is not always foolish. For instance, asking, "Where was God when my wife died?" can lead a grieving husband to Scripture's rich store of laments. On the other hand, contentiously demanding an answer to "Where are the dinosaurs in the Bible?" might rightfully need redirection or honing.

How do we know when "Where are the dinosaurs?" is the wrong question to ask? Does Scripture have nothing to say about the dinosaurs? Yes, it must have something to say about the fossilized history beneath our feet. But we may need to restrain or retrain our questions where the biblical authors are silent so we can instead hear what they are speaking loudly.

Lucy in the ground with baboon bones. In stories about both creation and evolution, mixtures of mythologies and causal explanations abound. Fictionalized men and women stand in for whole populations.

Even the renowned three-million-year-old "Lucy" (*Australopithecus afarensis*) is a fictional character amalgamated from a collection of scattered bones presumed to belong mostly to one prehistoric individual. I say *mostly* because her knee bone was discovered a year prior to the rest of her skeleton and over a mile away from the partial bone collection we have come to know as Lucy. In 2015, scientists realized that the Lucy skeleton also included an extinct baboon vertebra that had been wrongly included in Lucy's reconstructed spine.[4] Lucy, then, is the fictionalized character built around the conceptualized skeleton based on the 40 percent of a skeleton that was found roughly in the same area and including various bones from other things and places scattered across a kilometer radius.

Adam, a name conjured from the title in Hebrew (*ha'adam*), means something akin to "the dirtling" or "the earthling" in English. "The dirtling" (*ha'adam*) is his punny title precisely because he was taken from "the dirt" (*ha'adamah*). Like Lucy, Adam often appears in artists' fictionalized renditions along with imagined gardens and his "strong ally" later named Eve.[5] Dozens of sermons, church dramas, and children's services create new fictions every week of the first couple. They are depicted as arguing over fruit or wandering conversationally in a perfect tropical garden—perfect by the standards of whatever community depicts it.

But the creation stories in Genesis depict a "famously laconic" creation— it is too short by any standard.[6] Our unacceptably short biblical stories about

---

[4]Donald Johanson and Maitland Edey, *Lucy: The Beginnings of Humankind* (New York: Simon & Schuster, 1981), 159; Colin Barras, "Baboon Bone Found in Famous Lucy Skeleton," *New Scientist*, April 10, 2015, www.newscientist.com/article/dn27325-baboon-bone-found-in-famous-lucy -skeleton/; Marc R. Meyer et al., "Lucy's Back: Reassessment of Fossils Associated with the A.L. 288-1 Vertebral Column," *Journal of Human Evolution* 85 (2015): 174-80.
[5]Leslie Bustard gave me the phrase "strong ally" as a good-enough translation for *ezer kenegdo* (often translated as "helper fit for him" or "helpmeet"). I think it translates the situation quite well.
[6]"Biblical narrative is famously laconic." Robert Alter, *The Art of Bible Translation* (Princeton, NJ: Princeton University Press, 2019), 41.

creation bear such scant details that we often find ourselves telling stories *about* creation that are not *in* the biblical creation stories.[7]

Yet, stories of Adam and Lucy demonstrate that we help others understand our ideas about natural history by telling good-enough stories that generalize beyond the data. By good enough, I mean that they accomplish the task at which they aim without having to explain all the data. And like so many things, we create stories about creation for our own good or ill.

Joshua Swamidass's recent book, *The Genealogical Adam and Eve*, offers a great example of the former. He creatively offers narratives and explanations that thread the needle of current population-genetics research, the creation stories of evolutionary science, computational modeling, and the biblical accounts.[8] I hope more scholars follow his lead, and I will be returning to his model several more times.[9]

To see how the biblical authors also strategically employ narrative, poetry, and legal reasoning to explain reality, we must first value the intellectual world the biblical authors construct for guiding their endeavor. That is a big ask for some of us.

Let's be sober about this. Instead of entering expectantly into the Bible's intellectualism, we often try to relegate the texts of Scripture into some other category that allows us to get on with our task scientifically, or even spiritually. We find a way to swap out the biblical accounts for scientific mythologies until we get to "real history" within Genesis—usually starting with Abraham. Or, we create fictionalized accounts about Genesis to fill the gaping holes of the biblical creation accounts (e.g., depicting dinosaurs and humans living side by side). Both ways move the biblical texts to the side, presuming that they offer up only theological data (for the theistic evolutionist's task)

---

[7]William P. Brown includes Prov 8; Ps 104; Job 38–41; Eccles 1; and Is 40–55 in his examination of creation and science. Brown, *The Seven Pillars of Creation: The Bible, Science, and the Ecology of Wonder* (New York: Oxford University Press, 2010).

[8]S. Joshua Swamidass, *The Genealogical Adam and Eve: The Surprising Science of Universal Ancestry* (Downers Grove, IL: IVP Academic, 2019).

[9]Saying that scholars explain through fictionalized stories does not diminish their task, its explanatory power, or historical focus. Rather, this highlights the power of story to explain the most complex relationships and how to best understand them—in the sciences and Scriptures. Narrative, like deductive logic and poetry, constitutes a thought form: a way of thinking, clustering data, explaining, and arguing about the invisible features of the visible world. For a fuller account of narrative used as scientific and logical explanation, see Dru Johnson, *Biblical Philosophy: A Hebraic Approach to the Old and New Testaments* (New York: Cambridge University Press, 2021), 116-32.

or raw biblical data (for the creationist's so-called literal reading). But we are doing something better here, I think.

The biblical authors are reasoning with us. I want to wrestle with this claim: the biblical literature presents a coherent and sustained intellectual world for us to enter. This reasoning is meant both to construct and to delimit our philosophical (or theological) imaginations.[10]

In contrast, many Christian explanations that engage the biblical accounts end up treating the texts as either antiquated cosmic mythologies (how the world came to be), etiologies (causal explanations of our present world), functionalist explanations (how everything plays a role in ancient cosmology), or literalistic accounts that mix scientific reasoning and a flat reading of Genesis 1–2 (think Ken Ham's Answers in Genesis).

Paring the biblical creation texts down to functional, mythological, or etiological approaches does not sufficiently reflect the rigorous thinking presented by the biblical authors. And the literalistic readings do not go far enough, often not taking the literary aspects of intellectual literature seriously. All of these approaches could be partially mistaken in ways akin to my adolescent question about dinosaurs in the Bible. They create the potential to err by bringing their own problems to the biblical text and interrupting it with something that is problematic to them: typically, the many-splendored findings from the evolutionary sciences.

I want us to allow the biblical authors to set the terms of the discourse the way they saw fit. I want them to be the loudest voice when we consider the conceptual world of creation that they have constructed for their audiences, ancient and modern. Let us give *their* metaphysical frameworks, *their* values, and *their* priorities a hearing to see whether they are antiquated gobbledygook or something more sophisticated. When we do, we will see that the biblical authors argue more vociferously and with more sophistication than we might have imagined or fictionalized.

We will discover that the biblical literature suggests a form of natural selection, depending on what we mean by *natural*. We will have to puzzle that out later.

---

[10]On why the biblical literature represents a philosophical tradition, see Johnson, *Biblical Philosophy*; Yoram Hazony, *The Philosophy of Hebrew Scripture* (New York: Cambridge University Press, 2012).

We will also run the risk of having to say that the biblical authors might have been scientifically inaccurate about things.[11] Did the biblical authors appear to believe in a three-decker cosmos (i.e., heavens, earth, and the deeps) with a domed sky resting on a flattened land mass set on pillars surrounded by water above and below? Yes, most of them did seem to believe something like that.

Did they describe the cosmos in functional terms and trace causal explanations? At the least, yes, they did. Did their descriptions of creation provide a mythological account for later Hebrews (mythological in that it is an explanatory narrative about origins)? Yes, they did. Would the biblical authors believe that their accounts could hold up under modest scientific scrutiny if they were alive today? I think so, even if adjustments would have to be made to fit their understanding to current scientific understanding. Did the biblical authors write narratives in styles similar to what we have come to expect of historical or scientific explanations over the last century or so? Not so much.

I want us to pursue questions that I think are more interesting than these rather flatfooted questions for comparing creation and science. More than any of the above approaches on their own, the biblical authors constructed this literature to reason with their audiences about the nature of reality beyond the historical and concrete events they experienced, even beyond some present understandings we may have today.

As the simplest of examples about Iron Age Israelite concepts, the biblical authors make a steady distinction between reproductive anxieties and sexual generation. Sexual reproduction focuses on fertility and resident anxieties about producing children within one's life and to one's own benefit. Generation, on the other hand, looks over the horizon at the successive

---

[11]Inaccurate ideas do not entail that they were irrational ideas given the cosmological traditions and understanding of their age. For instance, Iron Age Hebrews appear to have held strong convictions about material cause-and-effect relationships but also included invisible forces in those mechanical calculations. They seem to have broken with Mesopotamian and Egyptian thought, being more rationally oriented (by our standards today) and demonstrating the earliest instances in literature of "critical intellectualism" and a "skeptical mood." Henri Frankfort et al., *The Intellectual Adventure of Ancient Man: An Essay on Speculative Thought in the Ancient Near East* (Chicago: University of Chicago Press, 1946), 234. See also Jan Dietrich, "Empiricism or Rationalism in the Hebrew Bible? Some Thoughts About Ancient Foxes and Hedgehogs," in *Sounding Sensory Profiles in the Ancient Near East*, ed. Annette Schellenberg and Thomas Kruger (Atlanta: SBL Press, 2019).

chain of reproduction. Generational anxieties focus on progeny in the land of the living, or what some might call genetic success. Most basically, reproduction relates to one's children, whereas generation relates to one's progeny.

And early on, Genesis disrupts the cultural flow of generational thinking by passing the covenants through the second-born instead of the first. This indicates that generational thinking is not merely about producing many lineages, but unexpected lineages. And sometimes these generated from the "weaker" sons, some who turned out to be scoundrels (e.g., Jacob over Esau). From Genesis's usurping of first-born son expectation to Paul's rhetoric about God choosing the weak and foolish things, biblical authors delight to disrupt our thinking about convention and strength.

This distinction between reproduction/children and generation/ descendants provides enough conceptual heft for later discussions about sexual behaviors driven by scarcity and fit. The biblical reproduction- generation distinction will be especially helpful in thinking about hominins who are capable of worrying about progeny beyond their immediate genera- tion.[12] Much more on this later.

Stories that argue with us. The Hebrews, like us, used stories to reason about their view of reality. Most of us grew up thinking that stories, on the page or the screen, were meant to entertain us. But stories can also make arguments. They can reason with us about the invisible features of the world. I might go further to say that narratives can uniquely inspect the unseen ways our world works, things not available to the naked eye and only grasped through insight. This unique ability of stories explains why filmmakers, scientists, philosophers, and sacred texts all use story to reason with us.

Pithy stories do some philosophical heavy lifting. They reason with us and initiate explanations. These argumentative stories, like scientific and biblical narratives, are rarely about what we can see with our eyes. Rather, they explore the invisible forces at work that make sense of what we can see. This is precisely what happens in most stories explaining the creation of humanity in the biblical literature and the sciences.

Concepts such as gravity and velocity must be understood apart from what is observed. Most (all?) concepts in the evolutionary sciences, including the

---

[12]Nota Bene: *Hominin* generally refers to humans, their extinct ancestors, and great apes.

concepts of Darwin's natural selection, cannot be observed with the senses. They must be understood in order to explain what can be observed.

The invisible concept of friendliness is one such example in evolutionary psychology. Brian Hare and Vanessa Woods's splendid book, *Survival of the Friendliest*, argues that natural selection ultimately favors friendliness. They define friendliness as the ability to share ideas and goals with others and accomplish them together.[13] When creatures begin demonstrating collaborative and beneficent traits toward others, there are physiological changes—to their teeth, face, torsos, and more—that can be predicted and traced. They call this self-domestication.

Hare and Woods then consider friendliness traits that allowed modern humans (*Homo sapiens sapiens*) to outlast other hominins such as Neanderthals (*Homo sapiens neanderthalensis*) through self-domestication. Modern humans did not necessarily outwit Neanderthals through their raw mental prowess. Rather, Hare and Woods believe that modern humans managed to outlast Neanderthals because of a new kind of social intelligence they conceptualize as friendliness.

But here is the point: Hare and Woods begin their book with a fictional story tracing friendliness through early hominins up to modern human beings. They then go on to examine twenty-first-century partisan politics in US Congress in light of their friendliness thesis. Ultimately, they conclude that fearmongering and dehumanizing of others in American partisan politics is antithetical to a natural selection that favors friendliness. They believe this all then yields a general moral truth: "Political rivals in a liberal democracy cannot afford to be enemies."[14]

This is all fascinating stuff, but I want us to notice how Hare and Woods themselves conceived of what they were doing in their book. They tell a story of how various versions of *Homo erectus* become various versions of *Homo sapiens* and how hunting, art, and communication allowed certain versions of hominins to flourish because of the invisible trait of friendliness.[15]

---

[13]Brian Hare and Vanessa Woods, *Survival of the Friendliest: Understanding Our Origins and Rediscovering Our Common Humanity* (New York: Random House, 2021).

[14]Hare and Woods, *Survival of the Friendliest*, xxx.

[15]Their fictionalized creation narrative imagines the cognitive challenges of hominins who first sailed the seas. They envisage those prehuman hominins venturing over the glassy horizon and how they had to prepare *today* for imagined situations in distant lands *tomorrow*. The authors

Hare and Woods label their own science-based narrative "a creation story" and insist that theirs is "not just another creation story." Their story should be the basis for shaping our morals and behaviors today: "It is a powerful tool that can help us short-circuit our tendency to dehumanize others. It is a warning and a reminder that to survive and thrive, we need to expand our definition of who belongs."[16] From their creation narrative they apply the principles seamlessly to political strife in twenty-first-century America.[17]

This is what I would like us to notice: they did not tell just any story of creation. Rather, they strategically aimed their good-enough story at invisible features of our world to foster our intellectual imaginations about how the world once *was*, now *is*, and *is to come*.

Like Hare and Woods's scientifically ensconced creation story, the biblical authors did not restrict themselves to what could be seen. They too attempted to give a good-enough story to elucidate the unseen features that then explain what we can observe today.

BRINGING IT ALL HOME

Did you catch what I tried to do above? In a very simple way, I attempted to find an intellectual common ground behind the storytelling in the creation stories of the Bible and scientific explanations (here *Survival of the Friendliest*). I did this because I want to compare them apples to apples. Both of them reason with us by presenting coherent intellectual worlds with robust explanations in the form of stories. They both rely on the conviction that a good-enough story can help us to see the unseen features that shape the world. And as a colleague reminded me, they do so "at least in part to convince us how to *live*! These are ethically freighted tales."[18]

---

even hint at the possible origins of abstract thinking and the intellectual development of imagination based on such fictions of possible hominin sea voyages.

[16]Hare and Woods, *Survival of the Friendliest*, xxx.

[17]The reality of friendliness as an organizing feature cannot be diminished by the fact that it cannot be seen. It must be understood through a collection of features shared between creatures that we can assess and label as "friendliness." So too with scientific metaphors such as electromagnetic *attraction*, *motivation* theory in social psychology, or the so-called *laws* of physics more generally. None of these metaphors—yes, they are all metaphors—can be directly observed by humans or instruments.

[18]Thanks to Michael Rhodes for pointing this out.

In the coming pages, I will trace the remarkable similarities between Darwin's version of natural selection and the biblical discourse on the same topics. Because Darwin's project only marked a beginning, I include early critiques of natural selection from within the sciences and some of the current ideas in the evolutionary sciences. (Of course, these surveys can only give glimpses into the most generally agreed-on ideas and critiques of natural selection. And the moving target that is "current evolutionary science" typically and only represents a slice of its theorists.) Alongside each of Darwin's aspects of natural selection—scarcity, fit, and sex—I trace how the biblical authors deal with the same group of selection pressures in their own way.[19]

I will suggest that the actual conflict between science and Scripture is not evolution versus creation. Rather, the conflict turns on how one answers this question about the present physical state of the universe: Is it now as it was in the beginning and ever more shall be?

But first, we dig deeper into Darwin's remarkable ideas.

---

[19]This book will deal with sensitive issues relating to violence and sexuality as seen in the evolutionary and biblical records. Readers should know these topics have been handled with care and respect, but also bluntly and descriptively when necessary. My hope is that we can learn from our dark human pasts so that we can confront present realities.

CHAPTER TWO

WHAT HATH DARWIN TO DO WITH SCRIPTURE?

"READ ARISTOTLE TO SEE whether any of my views are ancient."[1] Around 1851, while on the *HMS Beagle*, Charles Darwin penned this note to himself. Strangely, Darwin did not read Aristotle until just prior to his death in 1882, thirty years after scrawling this note. Even then, he read only an introduction to Aristotle and some passages from *Physics*.[2]

It should not surprise us that Darwin's burgeoning views on natural selection were met with "that sounds a bit like Aristotle" from his colleagues, presumably due to Aristotle's views on speciation.[3] The Authorized Version of the Bible was also listed among his books on board the *HMS Beagle*. He was, no doubt, familiar with the content of the Old and New Testaments even if he bristled at the idea of scientifically reasoning from them. In the end, we will see why Darwin would have done well to read more of Aristotle and also Scripture.

[1]Paul Barrett et al., *Charles Darwin's Notebooks, 1836–1844: Geology, Transmutation of Species, Metaphysical Enquires* (Ithaca, NY: Cornell University Press, 2009), 325.

[2]Allan Gotthelf, "Darwin on Aristotle," *Journal of the History of Biology* 32 (1999): 3-30, here 8.

[3]Indeed, William Ogle, who wrote the introduction and translation of Aristotle's *Physics* that Darwin later read, also sees a connection between the two men in their naturalist agendas. In a reply to Darwin's thank-you note for Ogle's translation, Ogle answers Darwin, "I feel some self-importance in thus being a kind of formal introducer of the father of naturalists [Aristotle] to his great modern successor [Darwin]." Even after Darwin read Aristotle, it is unclear to historians whether Darwin himself understood the connections that others might have seen between his work and the philosopher's. See Frederick Burkhardt et al., eds., *A Calendar of the Correspondence of Charles Darwin, 1821–1882, with Supplement* (Cambridge: Cambridge University Press, 1994), letter 13621, plate 2.

By looking at Darwin and Scripture side by side, we will see how their conceptions of nature, selection, and adaptation agree and depart from each other. We might have presumed the main conflict between Darwin and the Bible to be about random mutations and natural selection. But the deeper divergence actually lies in the uniquely Hebraic view of humans as moral actors who effect changes in the metaphysical orientation of the cosmos. That is a mouthful. Stated succinctly, morality affects physical reality. Thus, the conflict of visions includes more than superficially competing origins stories—evolutionary versus biblical.

However, as with his lifelong neglect of Aristotle, Darwin seems to have left the biblical literature largely unexamined.[4] His dismissive tone toward those ancient Semitic texts can be seen in his various correspondences (e.g., "I never expected to have a helping hand from the Old Testament").[5]

If Darwin had regarded those biblical texts as an intellectual peer, he would have discovered in his own Old Testament that his interests were quite ancient indeed, significantly more ancient and Asian than Aristotle's.

Predictably, the evolutionary sciences eventually broke from Darwin on key matters of natural selection because of later intellectual movements in biology and the computational sciences but also because of empirical research. As historians and philosophers of science have taught us, the models, metaphors, and conceptual worlds of every era of science inform what makes sense and to whom it makes sense.[6]

---

[4]His general pessimism toward the biblical creation appears several times in his correspondences. E.g., "I do not believe in metempsychosis nor in Genesis—& you are growing so orthodox, that you will end your days, I believe, in believing in the Tower of Babel." Charles Darwin to John Crawfurd, April 7, 1861, in *The Correspondence of Charles Darwin*, ed. Frederick Burkhardt et al. (New York: Cambridge University Press, 1994), 9:87.

[5]Darwin offers surprise at J. D. Hooker's use of Genesis to support his own view of natural selection that preserves a remnant, as did God in the flood. Darwin to Hooker, July 13, 1861. In his correspondences, it becomes clear that he views most attempts to concord "natural history" with Genesis as merely hawking superstitions and therefore largely unwanted. For example, "It's funny to see a man argue on the succession of animals from Noahs [*sic*] deluge; as God did not then wholly destroy man, probably he did not wholly destroy the races of other animals at each geological period! *I never expected to have a helping hand from the Old Testament.*" Charles Darwin to J. D. Hooker, July 13, 1861, in Burkhardt et al., *Correspondence of Charles Darwin*, 9:201, emphasis added.

[6]Inspired by Michael Polanyi, Thomas Kuhn went on to describe what happens when communities of scientist strike on an epistemological conflict: their skilled observations are not explained by the scientific theories du jour. Polanyi describes the structures and the revolutionary process a decade before Kuhn worked it out. See Kuhn, *The Structure of Scientific Revolutions*, 3rd ed. (Chicago: University of Chicago Press, 1996); Polanyi, *Personal Knowledge Towards a Post-critical*

I would like to think that if Darwin were reading this right now, he would think about what made sense to the Hebrews, to Aristotle, and to his colleagues. I think he would be interested. He might keep turning the pages, even if reluctantly.

DARWIN'S BIBLICAL IMPULSE

I will make the case in the coming chapters that an alliance of factors features centrally in *both* the biblical *and* Darwinist accounts of creation, some of which persist in evolutionary science. Moreover, this distinct discourse about the pressures of selection cannot be drawn from any other creation accounts in the ancient world. For most of us, that makes this conversation inherently interesting. If nothing else, our fascination with this Darwin-Bible parallel could be justified merely from the fact that it complicates that well-worn conflict narrative: science against religion. It is unexpected.

What makes the parallels interesting is not a bizarre causal link between them—as if Darwin unconsciously picked up his three pressure points of natural selection from his Old Testament and adapted them to constructs du jour. Rather, my interests are in Genesis's ability to critically engage and critique ideas such as Darwin's because Genesis speaks about the very same universal pressures on humanity: how to survive in this disordered cosmos. What is it about scarcity, fit, and propagation that makes them inherently useful for explanation in the Hebrew Bible, New Testament, and, later, Darwin's thinking?

APPLES AND ORANGES: APPROACHES TO DARWIN AND GENESIS

For a long time now, scholars have posed Genesis's creation accounts side by side with origin stories gleaned from evolutionary science—nothing new there. To say it again, we cannot forget that both the biblical and evolutionary accounts technically fit into the genre of mythos—narratives intended to explain the origins of matter, flora, and fauna. The genre of

---

Philosophy (Chicago: University of Chicago Press, 1974), 150-60; Struan Jacobs, "Michael Polanyi and Thomas Kuhn: Priority and Credit," *Tradition and Discovery: The Polanyi Society* 33, no. 2 (2006–2007): 26-36.

mythos does not deal with the question of historicity, only the aims of the story being told.

In her book *Evolution as a Religion*, skeptic philosopher Mary Midgley does not mince words on this front: "The theory of evolution . . . is . . . also a powerful folk-tale about human origins."[7] It is not as if the biblical account is a story and the accounts of evolutionary science represent "the facts."[8] They both employ narrative to collate the data for their rhetorical goals. These narratives are good enough, but never exhaustive in their details. Hence, Midgley deems them folktales (or mythos).

Whether these folktales are true to "natural history," in the ordinary sense of the term, is a matter for better minds. What matters most about this kind of mythos, whether biblical or scientific, are its rhetorical aims, even if set apart from the historical issues with which both folktales must inevitably deal. They both seek to explain a present reality by means of a dramatic story about a distinct and unfolding past. We will look at Darwin's stories of selection in the coming chapter. As we have already seen, the creation narratives of science continue to indicate how we should then live today.

Given the above parallels between Darwin's views and the rhetoric of Genesis, it might seem somewhat surprising that some Christian scholars are willing to set aside aspects of the biblical narratives as irrelevant at best or inconsequential at worst. However, the intellectual category in which they place the biblical texts informs why they think the biblical accounts are mythic *apples* to scientific *oranges*.

Genesis (and other biblical creation texts) tells of origins differently in tone and register from how we would today. The Bible's creation accounts do not address scientific concerns in the way we conceptualize them. The ancient Hebrews were using narrative somewhat differently than we typically think of narratives today. Many of these scholarly examinations of biblical and evolutionary approaches rightly point out those differences.

According to these scholars, biblical accounts sate the concerns of ancient audiences with mythos, while evolutionary science gives the most accurate

---

[7]Mary Midgley, *Evolution as a Religion: Strange Hopes and Strange Fears*, rev. ed. (New York: Routledge, 2002), 1.

[8]For more on how narratives are employed as logically rigorous forms of argument, see Dru Johnson, *Biblical Philosophy: A Hebraic Account of the Old and New Testaments* (New York: Cambridge University Press, 2021), 116-50.

account of natural history.[9] Many of these treatments of creation, especially by biblical scholars, lack attention to the intellectual world of Israel's Scripture.[10]

David Livingstone sums up one intellectual hurdle for scholars in the late nineteenth century that accounts for a version of this apples-to-oranges conflict:

> For long enough there had been theologians, particularly but not exclusively in Germany, who took a higher critical view of scripture. For them the Creation story of Genesis, with its depiction of Adam's creation, probation, and fall, were truly "mythological" and could therefore be sifted of all scientific content to leave behind the moral message as a sort of spiritual residue.[11]

It is incorrect to say that the origin stories in the evolutionary sciences are somehow equivalent to the biblical accounts in their aims to describe *what actually happened*. Scripture gives a particular story that leads to a particular people in history. Livingstone notes those who were attracted early on to Darwin's thought were also offended by the specificity of the man's "creation, probation, and fall" found in Genesis.

A universal and scientific account of humanity cannot have named people within it. Scientific creation accounts name the circumstances of creation, not the specific people or their lineages. Many increasingly found this one specific Hebraic story line untenable and unnecessarily binding—a specificity not required by a more comprehensive evolutionary account of how creatures came to exist in general.[12]

Another theological attempt reconciled Eden to the emerging evolutionism at the end of the nineteenth century by converting "Adam the specific person" into "Adam the character in a story" as a representative type standing in for all humankind.[13] This move attempts the same: to extricate

---

[9]Again, *mythos* is a technical term that does not necessarily mean "fictional" or "untrue" but rather a narrative that has a primary goal of explaining origins.

[10]There are exceptions, such as S. Joshua Swamidass's work, which uses the latest genetic and genealogical modeling to argue that the evolutionary and biblical accounts are simultaneously reconcilable. Swamidass, *The Genealogical Adam and Eve: The Surprising Science of Universal Ancestry* (Downers Grove, IL: IVP Academic, 2019). The observation about biblical scholars still applies, as Swamidass is a computational biologist, not a biblical scholar.

[11]David N. Livingstone, *Adam's Ancestors: Race, Religion, and the Politics of Human Origins* (Baltimore: Johns Hopkins University Press, 2008), 154-55.

[12]By Hebraic, I mean that it is Hebrew in language and origin.

[13]Charles Woodruff Shields argues for Adam as representative for all humanity, not an actual historical figure in a garden. See Livingstone, *Adam's Ancestors*, 156.

the actual origins story of humanity from the specific history of a man and woman as the direct genealogical ancestors of humanity. Again, the problem of the specificity of the Genesis narrative interferes. Or as N. T. Wright put it recently in terms of the new naturalism: "We had 'discovered' that a sharp division existed between the hard facts of this world, which does its own thing without divine intervention, and the vague fantasies of 'religion,' which were unprovable, unreliable, intolerant, and unhealthy."[14] An Eden that could be placed on a map with two named progenitors made Genesis a rather intolerable account for biblical scholars attempting to reconcile Genesis to evolutionary thinking.

AN APPLES-TO-APPLES APPROACH

Attempts to take the Bible's conceptual world seriously do not always make the proper points of contact with it. For example, in his essay on evolution and natural evils, Neil Messer wants to follow the biblical definition of what is considered "very good" in defining the problem of evil. Yet, he collapses the vast disparity between the "very good" world of Genesis and the one described by evolutionary biology, "a world deeply marked by scarcity, competition and violence—by the 'struggle for existence.'"[15] Messer cannot imagine how these are not contradictory accounts of our world—the world as it was, now is, and ever more shall be. Messer is then forced to puzzle out how our world, pockmarked with deficiency and violence, can be reckoned with the world of Genesis 1–2. The competition is between those two narratives, and the naivete of Genesis will have to lose.

I want to show that Genesis 1–11 precisely focuses our attention this way to explain "scarcity, competition and violence" in a world gone awry. This is the intellectual tradition played out across the biblical texts.

What Messer could not imagine is the uniquely Hebrew metaphysical imagination that Genesis seeks to construct. Hence, biblical authors from Genesis to Revelation argue about the disorientation of the cosmos and its rightful orientation. They do this by filling out our imaginations with a

---

[14]N. T. Wright, "Wouldn't You Love to Know? Towards a Christian View of Reality," September 1, 2016, http://ntwrightpage.com/2016/09/05/wouldnt-you-love-to-know-towards-a-christian-view-of-reality/.

[15]Neil Messer, "Natural Evil After Darwin," in *Theology After Darwin*, ed. Michael S. Northcott and R. J. Berry (Milton Keynes, UK: Paternoster, 2009), 139-54, here 148.

present reality unimaginable to most. Paul memorably depicts this imaginative hurdle as seeing "through a glass darkly" (1 Cor 13:12 KJV; or a "mirror dimly," ESV). At the end, we will have to spend some effort reimagining alongside the biblical authors to understand their metaphysical concepts, how they twist and turn.

If the Christian Scriptures explore the same or similar conceptual grounds to Darwin's natural selection and accounts from evolutionary science, then should their theology, begun in Genesis, not be offered a seat at the table of all Christian discussions of evolution?

To be clear, I am not evaluating whether some evolution narrative or a literalistic reading of Genesis is indeed *the Truth*. My goal is to offer an apples-to-apples account between two stories of origins and the conceptual worlds that support them. On the one hand, we have the ancient and specific Hebraic story of a particular man and woman in a geographically located garden, listening to a serpent, enacting his counsel, which then creates the disordered conditions of scarcity, mis-fit, and problematic genealogies. On the other, we have a generic humanity emerging and propagating from generic "natural" conditions of competition or cooperation in light of resources, which is determined by best genetic fit.

Can these two stories be reconciled? If not, why not? But even more fundamental than the stories they tell, at what points do their conceptual worlds conflict?

PART TWO

SCARCITY

CHAPTER THREE

DARWIN'S SCARCITY AND THE STRUGGLE FOR LIFE

YEARS AGO, MY FAMILY'S PLASTIC BIRD FEEDER became the setting of horrific prejudice and violence. It still scars my memories.

When our children were in primary school, we bought a hummingbird feeder—the apple-red kind with yellow florets that you might see hanging outside kitchen windows. We put ours on the front porch in the hope that we would see these iridescent marvels of flight from our front window. Our little children hunkered down at the window watching and waiting for hummingbirds.

I was excited too, but I warned the kids that we might never see a hummingbird and that insects might take over the feeder. Within a day or two, hummingbirds regularly swarmed the feeder. We could hear the vibrations of their fluttering wings through the open windows.

Then, a few days later, I heard the kids screaming from the couch. My daughters screeched, "The birds are killing each other!" I thought they were misinterpreting the scene until I saw it for myself. A few hummingbirds had ganged up to attack another. When the victim fell to the concrete, the perps pinned him down on the sidewalk while one of them began stabbing the pinned bird with his nectar-drinking beak. Thinking that I was watching a murder, I flung the front door open, hoping that would scare them off. It did not. I still had to manually shoo them off the bird on the ground. (It turns out that some hummingbirds use their beaks to stab others, specifically in the throat.)[1]

The children were crying as if they had just witnessed a gang execution, which we basically had. We took the feeder down and rinsed the nectar out of it. I remember thinking to myself, "I wish the birds knew that we had plenty of nectar for all to eat." Upon researching further, what we saw was likely a combination of territorialism and worries about food scarcity burned into these tiny winged motors. As one expert puts it, "The angriest hummingbirds may be aggressive well into the fall as they defend prime feeding spots in preparation for migration."[2] I'll say!

━ ━ ━

Scarcity often reigned in Darwin's thinking. Though Darwin clearly favored genetic fit as the most powerful force in natural selection, he also saw a synergistic relationship among an organism's environment, resources, and the oft-violent "struggle for life."[3] Due to limited resources, creatures compete, fight, and die by starvation, predation, dehydration, and all the interesting ways we die. Hence, the catchphrase "survival of the fittest" presumes the kinds of scarcity well known among our ancestors—a scarcity without recourse to community or government interventions.

Today, evolutionary biologists consider other factors such as mutation, genetic drift, and gene flow to be the most likely culprits for evolutionary change in a population. As the brutality of competition has been reassessed in the last generation of evolutionary thinking, cooperative models of evolution exhibit a similar impulse to the biblical authors. Scarcity-driven violence is a nonnecessary response in a world of want.

---

[1]Alejandro Rico-Guevara and Marcelo Araya-Salas, "Bills as Daggers? A Test for Sexually Dimorphic Weapons in a Lekking Hummingbird," *Behavioral Ecology* 26, no. 1 (January–February 2015): 21-29; James Gorman, "The Hummingbird as Warrior: Evolution of a Fierce and Furious Beak," *New York Times*, February 5, 2019, www.nytimes.com/2019/02/05/science/hummingbirds -science-take.html.

[2]Melissa Mayntz, "Hummingbird Behavior and Aggression: How to Tame Angry Hummingbird Behavior," *The Spruce*, April 14, 2022, www.thespruce.com/hummingbird-behavior-and -aggression-386447/.

[3]Today, most would argue that genetic mutation is the most significant force in evolutionary change, not natural selection. See Masatoshi Nei and Sudhir Kumar, *Molecular Evolution and Phylogenetics* (Oxford: Oxford University Press, 2000).

DARWIN'S STRUGGLE FOR LIFE

Darwin rarely let predation for resources slip from mind. What Disney later whitewashed in *The Lion King* as "the circle of life," Darwin painted with hues of destruction likely gleaned from Thomas Malthus's unsparing description of a "struggle for existence."[4] Darwin explores the nature of the struggle in the passage below, and he is worth reading at length here:

> All organic beings are exposed to severe competition. . . . Nothing is easier than to admit in words the truth of the universal struggle for life . . . than constantly to bear this conclusion in mind. Yet unless it be thoroughly engrained in the mind, the whole economy of nature, with every fact on distribution, rarity, abundance, extinction, and variation, will be dimly seen or quite misunderstood. We behold the face of nature bright with gladness, we often see superabundance of food; we do not see or we forget that the birds which are idly singing round us mostly live on insects or seeds, and are thus constantly destroying life; or we forget how largely these songsters, or their eggs, or their nestlings, are destroyed by birds and beasts of prey; we do not always bear in mind, that, though food may be now superabundant, it is not so at all seasons of each recurring year.[5]

Catchphrases of evolution drive the popular imaginations of Darwin's version of nature "red in tooth and claw."[6] However, we should not take it for granted that Darwin's version of natural selection by means of competition required such destroying. Darwin painted the canvas of a species's struggle more glorious than Lord Tennyson's oft-abused saying. It was a "struggle," but Darwin focused his affections on the outcome more than the melee. His was a teleological account as much as anything else: "Thus, from the war of nature, from famine and death, the most exalted object which we are capable of conceiving, namely, the production of the higher animals, directly follows."[7]

The mechanism, for Darwin, was just that: a perfunctory mode of tailoring and tailing off unfit species. In this new story about an ancient "tree

---

[4]Peter J. Bowler, "Malthus, Darwin, and the Concept of Struggle," *Journal of the History of Ideas* 37, no. 4 (October–December 1976): 631-50.

[5]Charles Darwin, *The Origin of Species, by Means of Natural Selection*, 6th ed. (London: John Murray, 1872), 49.

[6]Lord Alfred Tennyson, *In Memorium A. H. H.* (London: Edward Moxon, 1850), canto 56.

[7]Darwin, *Origin of Species*, 429.

of life," the branches that adapted lived to fight another day: "For as all organic beings are striving to seize on each place in the economy of nature, if any one species does not become modified and improved in a corresponding degree with its competitors, it will be exterminated." The environment can also act as the agent of selection. The extremes of scarcity and climate test an organism's fit and kill off those that do not. So, Darwin envisions climate as another set of pruning shears. "Periodical seasons of extreme cold or drought seem to be the most effective of all checks."[8]

This is not, for Darwin, a sad story of the natural world. Rather, he reminds the anxious reader in the final sentence of a chapter titled "Struggle for Existence" that "death is generally prompt, and that the vigorous, the healthy, and the happy survive and multiply."[9] I guess our hummingbird execution squad was not so bad after all.

In the evolutionary sciences, the seeming natural connection between scarcity and competition has come under repeated challenges. Beginning in the early twentieth century up to more recent feminist critiques, a simplistic account of scarcity that leads to competition now faces significant scrutiny within the evolutionary sciences.

For instance, Robert Ricklef confidently claims competition as fundamental for shaping adaptation. "A resource consumed by one individual can no longer be used by another." But Michael Gross and Mary Beth Averill push back that if this is so, "then almost any behavior can be so construed . . . [as] competitive."[10]

Despite the prominence of competition in his thinking, Darwin also puzzled over the probability of biological altruism, organisms sharing resources to their own disadvantage.[11] But his interest remained mostly in the civilizing role of social altruism in humans. Darwin marveled at "a savage,

---

[8]Darwin, *Origin of Species*, 80, 54.

[9]Darwin, *Origin of Species*, 61.

[10]Robert Ricklef, *The Economy of Nature* (Portland, OR: Chiron, 1976), 266, quoted in Michael Gross and Mary Beth Averill, "Evolution and Patriarchal Myths of Scarcity and Competition," in *Discovering Reality: Feminist Perspectives on Epistemology, Metaphysics, Methodology, and Philosophy of Science*, ed. Sandra Harding and Merrill B. Hintikka (New York: Kluwer Academic, 2004), 71-95, here 89.

[11]Samir Okasha, "Biological Altruism," in Stanford Encyclopedia of Philosophy, Summer 2020 ed., ed. Edward N. Zalta, https://plato.stanford.edu/archives/sum2020/entries/altruism -biological/.

who will sacrifice his life rather than betray his tribe."[12] He could not have been aware of how much those puzzling altruistic behaviors might have shaped his thinking on the matter because he did not know how widespread they were.

It is not just humans who sacrifice for others but even trees and chimpanzees. Darwin could not have known, might not have been able to even speculate, of the social networks that share requests for nutrients in the forest in lean times. Trees, as one example, not only use fungal networks to communicate needs between them down in the forest floor, but do so between different species.[13] More on trees in chapter eight, but for now we can see that Darwin's puzzlement over savage men self-sacrificing when they did not have to might be overwhelmed by recent findings.

Cooperation. Competition, in harsh environs low on resources, generates violence. That *was* the thinking. The need to include cooperation highlights an advance in evolutionary thought: fittedness *to each other* has deeply embedded prosocial and reciprocal features.

The idea that cooperation might equally fund evolution goes back to at least the 1930s but gained momentum by midcentury with the work of Ashley Montagu, who not-so-subtly labeled the focus on competition "the Darwinian Fallacy."[14] Feminist theorists followed and dogged the dogma of Darwin's views regarding scarcity-fueled competition.

These critiques aimed at the undemonstrated and universally presumed metaphor of nature as a field of battle. Feminists argued that this model stems from an androcentric portrayal of nature "red in tooth and claw." Mostly male biologists throughout history have produced unsurprisingly male visions of nature. Thus, "nature, as depicted in biological science, is a man's world." This male emphasis on "parsimony may derive largely from

---

[12]Charles Darwin, *The Descent of Man, and In Relation to Sex* (Princeton, NJ: Princeton University Press, 1981), 165.

[13]Diane Toomey, "Exploring How and Why Trees 'Talk' to Each Other," Yale School of Forestry and Environmental Studies, September 1, 2016, https://e360.yale.edu/features/exploring _how_and_why_trees_talk_to_each_other/.

[14]The earliest assertion of the fundamental role of cooperation I could find was William Patten, "Coöperation as a Factor in Evolution," *Proceedings of the American Philosophical Society* 55, no. 7 (1916): 503-32, followed by Frank Robotka, "A Theory of Cooperation," *Journal of Farm Economics* 29, no. 1 (February 1947): 94-114. Following in 1950 was Ashley Montagu, "Social Instincts," *Scientific American* 182, no. 4 (April 1950): 54-57; and Montagu, *Darwin: Competition and Cooperation* (Westport, CT: Greenwood, 1952).

male socialization to strive against others and to manipulate nature in the world of work; and it may little correspond with women's traditional experience."[15] In other words, the evolutionary constructions of two sexes, specifically with aggressive and stronger males in hominins, might have also helped to shape the theories of evolution.

Notably, male evolutionary scientists were the first to critique Darwin's competition thesis. J. B. S. Haldane argued in the 1930s for a softer and kinder selection, at least among animals with cognition, "In so far as it makes for the survival of one's descendants and near relations, altruistic behavior is a kind of Darwinian fitness, and may be expected to spread as a result of natural selection."[16] Ashley Montagu followed suit in the 1950s: "Combative competition and struggle *may* in some cases lead to one type leaving a greater progeny than another, but it does not necessarily follow that such processes will always do so . . . *combative competition and struggle are not by any means necessary parts of the modern theory of natural selection.*" Montagu then authored a full book on the topic two years later: *Darwin: Competition and Cooperation*, which he wryly claims in the preface "should be subtitled *The Darwinian Fallacy.*"[17]

Based on new research on cooperation within evolution, skeptic philosophers such as Mary Midgley, biologists/mathematicians such as Martin Nowak, and theologians such as Sarah Coakley similarly argue against social Darwinism.[18] Most basically, if cooperation were central to selection, then natural selection such as Darwin's struggle for life simply cannot explain all the data. Many (maybe even most) scientists and theorists today see cooperation differently from Darwin. Nowak writes, "Thus, we might add 'natural cooperation' as a third fundamental principle of evolution beside mutation and natural selection."[19]

---

[15]Gross and Averill, "Evolution and Patriarchal Myths," 71, 76.

[16]J. B. S. Haldane, *The Causes of Evolution* (New York: Longmans, 1935), 131.

[17]Montagu, *Darwin: Competition and Cooperation*, 104, emphasis added. The proposed subtitle has two targets: Darwin's own view of natural selection as driven exclusively by competition, and those who have misunderstood Darwinism.

[18]Mary Midgley, *Evolution as a Religion: Strange Hopes and Strange Fears* (New York: Routledge, 1985), 6-8; Martin A. Nowak, "Five Rules for the Evolution of Cooperation," *Science*, December 8, 2006, 1560-63; Sarah Coakley, *Sacrifice Regained: Reconsidering the Rationality of Religious Belief* (New York: Cambridge University Press, 2012), 22-27.

[19]Nowak, "Five Rules for the Evolution."

The burgeoning biological case for cooperation demands some explanation as well. Brian Hare and Vanessa Woods argue suggestively in *Survival of the Friendliest* that the physiology of domestication dovetails with cooperative features in mammals apart from human intervention.[20] Not only is domestication a human-driven endeavor, but *Homo sapiens* seem to have gone through a biological selection for friendliness uncoerced by other beings. The so-called natural world created the conditions for which humans were domesticated through selection of friendliness features. The selection of these physiological and psychosocial features then further shaped human physiology and bent them toward socialized cooperation.

In short, although the premise that *scarcity entails competition in a struggle for life* became standard for the initial stages of evolutionary thinking, questions arose about the necessity of competition in the struggle. Struggle does not necessitate competition, since struggle and biological mechanisms can be shown to produce teamwork, from single-celled organisms to trees to hominins. Gross and Averill conclude that the lack of evidence for competitively driven adaptations creates "no good reason to attribute such characteristics to a 'competition' to develop protective attributes." They continue, "Why not see nature as bounteous, rather than parsimonious, and admit that opportunity and cooperation are more likely to abet novelty, innovation, and creation than are struggle and competition. Evolution in this perspective can be seen not as a constant struggle for occupation and control of territory but as a successive opening of opportunities."[21]

Though this turn to cooperation in evolutionary thinking is provocative and helpful, one wonders how much we can imagine enlarging the pie of resources before it must inevitably be divided—and division will mean life for some but not all. Even today, division of resources generally compels struggle that turns to violence among humans and against their habitats. Studies suggest that even something as basic as the threat of diminishing water supplies can explain much of subnational, national, and international politics, including wars.[22]

---

[20]Brian Hare and Vanessa Woods, *Survival of the Friendliest: Understanding Our Origins and Rediscovering Our Common Humanity* (New York: Random House, 2021), 99.

[21]Gross and Averill, "Evolution and Patriarchal Myths," 85.

[22]Scott M. Moore, *Subnational Hydropolitics: Conflict, Cooperation, and Institution-Building in Shared River Basins* (New York: Oxford University Press, 2018).

It is too soon to place our bets on either a competition- or cooperation-driven origins story in the so-called natural history. Some blend of both will likely rule future thinking in the evolutionary sciences. But notice that both explanations struggle to control a narrative, neither of which finds unyielding foundations in the empirical sciences. Yet, both speak crucially to the nature of human morality.

Montagu balked at what he considered Darwin's momentous error, fixing the whole of natural selection in the evils of competition to the death. "Darwin helped to establish such seeming paradoxes as that good could flow from evil, and that in the biological sense such evils were really good."[23] Feminist critiques suppose that male-driven explanations will unnecessarily read adversarial survival into the map of biological observations when it need not be the case. At the present, these still resemble what Midgley called origins "folktales" with some mathematical modeling behind them.

The feminist critique asks for the possibility of a different imagination surrounding natural selection. It allows the female experience to participate in shaping and forming concepts and possibilities of selection. The species adapts through cooperative means in abundance—a sentiment to which Darwin might have been amenable.[24] In all cases, the conception of nature determines what kind of "natural" selection is then projected back onto natural history. Can nature be, for the evolutionary sciences, a cooperative evolutionary environment from cells to souls, or does the red claw of competition always loom along the way?

Violence and scarcity. Though violence is not a necessary outcome of scarcity, some evidence suggests that violence often accompanies scarcity, specifically when it comes to hominins. To be clear, the idea here is *not* that evolutionary processes create violence; rather, scarcity can.[25] Scarcity can

---

[23]Montagu, *Darwin: Competition and Cooperation*, 99.

[24]Of course, Darwin himself suggested precisely this in *The Descent of Man*. Darwin points to the tendency of males to have showy features to impress the females in mating rituals, which ends in a sexual propagation of the male's genes. If male features win them sexual mates, then the preference of females for those features determines part of the evolutionary path for that species. See Darwin, *Descent of Man*, 198-204.

[25]"Evolution therefore cannot purposefully react to possible environmental changes. Merely the competition for resources, the struggle for existence, ultimately allows those individuals who have adapted best to increase in disproportionate numbers over the long term." By increase of species, they mean that some will die while others proliferate. Ina Wunn and Davina Grojnowski, *Ancestors, Territoriality, and Gods: A Natural History of Religion* (New York: Springer, 2016), 29.

also create cooperation. So no simple and singular connection can be drawn between scarcity/plenty on the one hand and competition/cooperation on the other.

Speaking of animal violence, humans are particularly violent creatures. When fit into a broad survey of animal violence, humans "are an average member of an especially violent group of mammals, and we've managed to curb our ancestry." One survey reviewed archaeological and epidemiological data on cause of death from 50,000 BCE to the present in order to suggest that "at the origin of *Homo sapiens*, we were six times more lethally violent than the average mammal, but about as violent as expected for a primate."[26]

There may be some modest evidence to loosely support the view that hominins are particularly violent in times of resource scarcity. Lester Brown and Gail Finsterbusch recall the recorded history of the savage effects of ancient famines. When things got desperate, even highly stratified cultures would "sell children to obtain money for food; and resort to suicide, murder, and cannibalism."[27] Families drowned themselves in the Tiber in Rome (436 BCE) to alleviate starvation. Cannibalism factored heavily in India, Europe, and the colonized Americas—for example, Jamestown—in times of severe famine.[28] Accounts of murder for the sake of cannibalism during famine even come from Iron Age Israel (2 Kings 6:24-31).

Famine offers an extreme example, but it does not represent the typical mode of violence from scarcity pictured in Darwin's view. Scarcity allows specific members of a species with specific genetic advantages to acquire the scarce resources. Famine, on the other hand, leaves most organisms resourceless.

A recent survey of California's prehistoric graves found a correspondence between the climate's lean years and the number of deaths by blunt-force trauma to the head or pointy objects, for example, spearheads. Scarcity

[26]Ed Yong, "Humans: Unusually Murderous Mammals, Typically Murderous Primates," *The Atlantic*, September 28, 2016, www.theatlantic.com/science/archive/2016/09/humans-are-unusually-violent-mammals-but-averagely-violent-primates/501935/; José María Gómez et al., "The Phylogenetic Roots of Human Lethal Violence," *Nature*, October 13, 2016, 233-37.

[27]Lester R. Brown and Gail W. Finsterbusch, *Man and His Environment: Food* (New York: Harper & Row, 1972), 2.

[28]Nicholas Wade, "Girl's Bones Bear Signs of Cannibalism by Starving Virginia Colonists," *New York Times*, May 1, 2013, www.nytimes.com/2013/05/02/science/evidence-of-cannibalism-found-at-jamestown-site.html.

correlated with violent deaths more than anything else. Mark Allen and colleagues report, "Sharp force trauma, the most common form of violence in the record, is better predicted by resource scarcity than relative sociopolitical complexity."[29] Who is to say why? Violence "associated with" scarcity seems to be the best we can say, and it is the same model often used in Scripture. Correlations between violence and resource anxieties do not entail causation, but repeated pairing of the two raises the question: How does scarcity contribute to violence, and if so, does it do so by necessity?

Some studies suggest that knowledge based in experience of scarcity has adaptive benefits in our cognitive abilities. Kelly Goldsmith and colleagues think scarcity could make us more adaptable: "Reminders of resource scarcity elicit cognitive flexibility, in service of the desire to advance one's own welfare." Others found that personal experience of scarcity adversely distorted decision-making in adults.[30]

Bonobo apes are a notable exception to the above and give us a living model of nonviolent primates—the only apes that do not hunt or kill other monkeys. Many have heard of the bonobo chimpanzees because they are humanity's closest genetic relative and they use wildly promiscuous sexual gratification (by hominin standards) to navigate social situations.

Relevantly, bonobos also go out of their way to share food when it is of no advantage to them and will help a stranger bonobo get something the stranger wants—a feature commonly seen in human children that fades as humans get older.[31]

Key to this discussion of bonobo prosocial behaviors and aversion to violence is one other fact: they live in a food-abundant forest in the Democratic Republic of Congo. They do not compete for food, and some believe that creates the conditions for their peaceable behaviors.

---

[29]Mark W. Allen et al., "Resource Scarcity Drives Lethal Aggression Among Prehistoric Hunter-Gatherers in Central California," *Proceedings of the National Academy of Sciences* 113 (October 2016): 12120-25.

[30]Kelly Goldsmith, Caroline Roux, and Anne Wilson, "Acting on Information: Reminders of Resource Scarcity Promote Adaptive Behavior and Flexible Thinking," *Advances in Consumer Research* 45 (2017): 253-57; Chiraag Mittal and Vladas Griskevicius, "Early-Life Scarcity, Life Expectancy, and Decision-Making," *Advances in Consumer Research* 45 (2017): 253-57. These authors did not look at scarcity in the context of a "struggle for life," so they can only be suggestive here.

[31]Frans B. M. De Waal, "Bonobo Sex and Society," *Scientific American*, June 1, 2006, www.scientific american.com/article/bonobo-sex-and-society-2006-06/.

CONCLUSIONS

For Darwin, scarcity begets a struggle for life. So, scarcity plays a crucial role in pruning the tree of life: "The amount of food for each species, of course, gives the extreme limit to which each can increase; but very frequently it is not the obtaining food, but the serving as prey to other animals, which determines the average number of a species."[32]

Whether or not empirical studies can adequately show that scarcity necessarily leads to violence and under what conditions, the twinning of scarcity and violence is not irrational. The two have commonly been considered together, both by Darwin and advocates of the cooperationist models of natural selection. Thus, when we see the biblical authors making the same associations, they are not speaking out of turn. They are making the same connections that humans have always made between scarcity and violence. But like those in the cooperative camp of natural selection, they also make the claim that scarcity does not necessarily require violence in response.

So how do the biblical authors map out their reasons for scarcity and its relation to violence?

---

[32]Darwin, *Origin of Species*, 53.

CHAPTER FOUR

SCRIPTURE'S INTELLECTUAL WORLD

IN THE FILM *Willy Wonka and the Chocolate Factory,* Wonka croons about a world of "pure imagination": "Travelling in the world of my creation, what we'll see will defy explanation." To our horror and glee, we slowly realize Wonka has rigged the factory tour as a series of psychopathic morality tests for the true golden-ticket holder. It raises the question: What is Wonka's "pure imagination"? He seems to mean something like "unrestrained innovation." That is one way we often think about imagination: some free-ranging faculty in our mind that allows us to bring new things into existence. There are, of course, other ways to conceptualize our imagination.

We will examine scarcity in Scripture in the next chapter. But before we can do that, we must understand that the biblical authors want us to imagine something we *cannot* bring into existence: a new metaphysical orientation of the presently existing world. So they construct an intellectual world in which we can imagine but not comprehend the way the cosmos once was, based on how we experience it now, in order to grasp what it will be one day.

The *Gloria Patri*—"as it was in the beginning, is now, and ever shall be"—celebrates the surety of God over time. But the biblical authors reason with us about the disoriented nature of creation itself, what we might call the *Gloria Terra*: "As it was in the beginning, *no longer is*, but one day shall be."

It once was (Eden) → no longer is (present) → it shall be (new heavens / new earth)

Such reasoning requires our imagination, even if through a glass darkly. This imagination is not merely free ranging, nor necessarily creative, not in the colloquial sense, at least. Rather, it extends from a thoroughgoing grasp of a presently disoriented cosmos.

Those ancient Hebrew authors are not the only ones requiring us to use our imaginations. The evolutionary sciences also ask us to imagine a world different from the one we know today. But that ancient world was only different in the kinds of creatures that existed, not in the so-called laws of nature. In the evolutionary world, the metaphysical nature of the universe remains unchanged. The laws of thermodynamics, gravity, electromagnetism, and the like persist. This means biology plays in the same realm of physics as it always has. We might say that scientists often think in terms of a *Gloria Physicis* (to play off the *Gloria Patri*):

> As it was in the beginning, *has always been*, and ever shall be.

My task here is straightforward: to consider how Darwin and recent evolutionary scientists portray the fundamental factors of human origins alongside how the biblical authors depict the same. Evolutionary science has gone on to outline several other factors aside from natural selection (e.g., gene flow, gene drift, mutation, and cooperative selection). I will focus on natural selection and touch on mutation for one simple reason: the biblical authors develop a thick discourse from Genesis about the topics of natural selection that touches on the nature of mutation in a way that is historically unique to them.[1]

By now, you can probably recite the germ for this present exploration: the biblical authors had their own intellectual world, which overlaps significantly with the topics of natural selection identified by Darwin. Instead of answering why these aspects of evolution feature so prominently in both the biblical accounts and Darwin's work, here I will demonstrate that they do indeed play starring and sometimes supporting roles in both origin stories: Darwin's and Israel's.[2]

---

[1] There are, of course, other creation accounts in the Hebrew Scriptures. William P. Brown includes Prov 8; Ps 104; Job 38–41; Eccles 1; and Is 40–55 in his examination of creation and science. Brown, *The Seven Pillars of Creation: The Bible, Science, and the Ecology of Wonder* (New York: Oxford University Press, 2010). I would see these as largely complimentary accounts, but I do not believe Brown's argument impedes anything I am arguing for here.

[2] Why both cover similar biological themes, I do not know. This raises a question that I cannot adequately answer: Why do scarcity, our physiological relation to environment, and the hope of

If Scripture teaches adequately and accurately about scarcity, fit, and propagation, then we might think that Darwin's views reflect something profoundly ancient by homing in on that same collection of features. Scripture's focus on the pressures of natural selection makes Darwin's observations feel like he is reaching back in time to touch something universal to the human experience, like touching the thirty-thousand-year-old handprints painted on the walls of the cave at Chauvet in France.

Israel's story is one of survival. Darwin's story deserves to be put in conversation with it on these fronts, even if the two cannot always be reconciled. I invite the reader to think with me how the biblical authors, if brought up to speed on scientific findings and theorizing, might have affirmed and critiqued Darwin and evolutionary science as conversation partners.

UNIQUELY HEBRAIC PARALLELS

Christian apologists and many biblical scholars often claim that the creation accounts of Genesis are different from all the other religious traditions. Though not obvious in Genesis, the Hebrew creation accounts are supposedly monotheistic (one god) instead of henotheistic (one head god among many). The one god of Genesis creates by speaking rather than through violence and turmoil, with a notable lack of churlish gods stirring up dissent. But I am not thinking about those kinds of arguments for its uniqueness. My claim centers on the conceptual economy of Genesis, how it thinks about the nature of *nature* and the relations inherent within it.

Suppose Darwin had read even further afield from Aristotle to survey the origin stories of Mesopotamia, India, China, Greece, and Egypt. Darwin would not have found this triple entanglement of want, aptness to environment, and propagation of species. Only in the Hebraic origin stories do we find these three concepts tightly interwoven and central to the drama of the creation narrative.

Mesopotamian creation dramas focus their energies on theogony: the creation of the gods. Humans come as a secondary thought, most often to

---

sexual propagation feature so prominently and uniquely in both the biblical creation and Darwin's schema of adaptation? And why do they not appear in other ancient creation accounts? I will set those questions aside and present instead the biblical authors' perspectives and prescriptions on scarcity, fit, and propagation, as well as Darwin's and contemporary debates.

relieve the gods themselves from having to dig canals and to build houses (temples) for them, and, most importantly, to offer oblations. When the humans become bothersome to the gods—too noisy, in one account—the gods flood the land in order to kill them off for quiet's sake.[3]

Stories from the Indian subcontinent contain their own reasons for conceiving of creation quite differently from those stories of the ancient Near East. Outside the Indian Manu tradition, where one Noah-like character builds a boat to survive a divine flood, Brahmanism and its Buddhist heirs focus their creation accounts on the cosmic scope.[4] The fabrication of matter and *ātman* (being) is central; humanity less so. Hymns of human creation are terse and mainly reveal that humans are made to offer oblations to the gods.[5] Some Buddhist philosophers, such as Vasubandhu (ca. fifth century BCE), even lodge quite technical arguments against the notion of a creator god as absurd based on principles of change and causation.[6]

The Chinese creation tales are more recent (ca. third century CE) and, like the ancient Near East, largely unconcerned with a drama of humans at creation. The most noteworthy version is the making of the universe from P'an Ku's divine-but-decomposing body, where humans feature as the mites that are nibbling on his rotting corpse.[7] That is about all we get.

Egyptian creation stories, like Mesopotamia's, focus mostly on cosmogony and theogony—creation of the cosmos and gods respectively. However, one story does resonate with the biblical depiction of God forming the man from the dirt near Eden (Gen 2). In Egypt, we find a repeated depiction of Khnum, the ram-headed god, creating humans and animals on a potter's wheel. The goddess Heqet sits opposite of Khnum offering the "breath of life," symbolized by the ankh in the nose of the recently formed creatures, which animates them. One such text reads, "He made the breath of life for their nostrils. They are his images which came forth from his body. He shines in the sky at their desire."[8] Other linguistic connections between

---

[3]Epic of Gilgamesh, tablet XI.

[4]E.g., the cosmic egg account of creation in the Rig Veda, book 10, CXXI.

[5]Rig Veda, book 10, CXXIX-CXXX.

[6]Jonathan C. Gold, "Vasubandhu," in Stanford Encyclopedia of Philosophy, Summer 2018 ed., ed. Edward N. Zalta, https://plato.stanford.edu/archives/sum2018/entries/vasubandhu/.

[7]See the P'an Ku (a.k.a. Pangu) and Nü Kua accounts.

[8]James K. Hoffmeier, "Some Thoughts on Genesis 1 & 2 and Egyptian Cosmology," *Journal of the Ancient Near Eastern Society* 15 (1983): 39-49.

these Egyptian divine-potter texts and Genesis 2 make the Egyptian creation narratives the closest kin with what we find in Genesis 1–2.[9]

Despite those similarities, Genesis 1 portrays the making of the cosmos and moves quickly to the formation of habitable regions and the creatures therein.

It is only in the Hebrew creation that humans are then commissioned as regents over the inhabited realms. Only the Hebraic account then reworks the commissioning of humanity (Gen 1:1–2:4)—their call to propagate and manage resources—into a story demonstrating their distinct fittedness to earth's greenscape and to each other (e.g., Gen 2:4-25). This story even commissions the animals to sexually reproduce in order to fill the habitable lands, skies, and seas respectively.

Israel's creation narratives incomparably weave together these two accounts with an ensuing genealogical tree of life following the stories of Semitic (literally: descended from Shem) tribes and clans. The biblical authors weave all these concepts into an inextricably bound chorus of stories.[10] Of the many people groups listed and presumed in the Hebrew Bible, only a few of Shem's descending clans will be fit to survive the trials of their covenant history with God.

If Darwin had reread all of these ancient origin folktales and philosophers, he would have found out that his views did have some ancient roots in Aristotle, *but even more so in the Hebrew Bible.* The remarkable similarities between the obsessions of the biblical authors and insights of Charles Darwin deserve consideration. I will say more in the coming pages about the overlap among the conceptual worlds of Scripture, Darwin, and evolutionary science. But it is worth noting how striking this is, that the biblical authors worked and reworked these concepts across the cosmic drama of Christian Scripture, and no one else seems to have done the same.

---

[9] See Hoffmeier, "Some Thoughts on Genesis 1 & 2," for the other linguistic connections.

[10] "Analysis reveals that the apparently 'artless' story of man and woman in the garden of Eden has in fact structures and intricate patterns of organization that involve even minor details of the text. Moreover, the patterns so interlock that the deletion of any part of the text (except, perhaps, 2:10b-14) would have significant repercussions for the whole passage." Jerome T. Walsh, "Genesis 2:4b–3:24: A Synchronic Approach," in *I Studied Inscriptions from Before the Flood: Ancient Near Eastern, Literary, and Linguistic Approaches to Genesis 1–11,* ed. Richard S. Hess and David Toshio Tsumura (Winona Lake, IN: Eisenbrauns, 1994), 362-82, here 375.

Like many of his day until now, Darwin appears to have treated Scripture as oracle, not an intellectual tradition. Without sufficient reason for doing so, we cannot fault him for not considering its unique overlap with his own work. However, we should certainly reflect on the reasons some still do not consider the Scripture as a formative intellectual tradition.

TERMS, DAMNED TERMS, AND CONCEPTUAL WORLDS

Of course, Darwin's conceptual world that he himself develops does not define all evolutionary thought or the sciences in his day or ours. In referencing Darwin, I am talking about the man himself. I do not intend to reference "Darwinism" as an atheistic ideology, atheistic views of biological processes, or any other such views.[11]

I mean only to engage Darwin in his writings and with subsequent scientists and philosophers who have adopted aspects of Darwin's views where they are empirically testable. I will refer to Darwin when speaking of what he has written.

I use the term "evolutionary science" when dealing with empirical evidence of the theory and "evolutionary theory" when referring to a collection of ideas on biological change through adaptation. All terms in this arena are plastic and have appropriated various meanings by different thinkers and groups.

Despite this variation, often driven by new understandings of biological systems not available to Darwin (e.g., gene flow, genetic drift, point mutation, etc.), there is a conceptual world of evolutionary thought. Just like the conceptual world of the biblical authors, I will engage the world or universe of discourse of evolutionary science when discussing biological adaptation. The interdependent and abstract notions within that conceptual world are what I am calling evolutionary theory.

Hence, ideas within evolutionary theory can be in direct conflict with one another. For instance, cooperative versus competitive views of natural

---

[11]The words used here participate in a land-mined matrix of creationist and evolutionary terms. Many of us bring a discourse of trauma to the discussion, alienated or embroiled by strategic use of rhetoric to persuade anyone who will listen. Those triggering terms will be part of my examination too. I will do my best to not embroil.

selection are both within the conceptual world of evolution and depend on other concepts within that world to make sense (e.g., environmental fit, sociological models, reactions to scarcity, etc.). However, some theorists have pitted cooperation against competition as fundamental for natural selection.

This book is about the conceptual worlds of evolution and the biblical authors. The focus is on how these two worlds conceive of humans in relation to their environment. What is the point of this? So that I can offer theological suggestions that stem from putting these two worlds in conversation.

WHAT I AM AND AM NOT ATTEMPTING

First, I am not aiming to address the natural history of biological life on earth. I will not tackle the record of animals that seem—by our best assessments—to have lived and died millions of years ago up to the present. I focus where the origin stories of evolutionary science and Scripture focus: on the successful and propagating branches of biology, specifically human history.

Second, I am not engaging arguments of historicity regarding the garden at Eden or the first couple (popularly known as Adam and Eve). The historic value of those biblical narratives is separable from this task, though less so for forming theology based on this study.[12]

Third, I am not attempting to prove or disprove evolutionary science or biblical history. The goal is to compare how two speculative worlds—ancient Hebraic and modern scientific—conceive of and argue about their shared concepts.

Fourth, though I will focus on humanity in both evolutionary models and biblical discourse, humans are never separable from flora and fauna. In both biblical and common thinking today, humans are held specially, though not solely, accountable for their actions (e.g., who gets blamed for air pollution, methane-producing livestock or their owners?). The Scriptures go much further. Humans and animals alike are commissioned by God to be sexually

---

[12]Plausible genetic arguments about Eden's historicity already exist. I will address this model later. See S. Joshua Swamidass, *The Genealogical Adam and Eve: The Surprising Science of Universal Ancestry* (Downers Grove, IL: IVP Academic, 2019).

fruitful (Gen 1:22), they die side by side in the flood (Gen 7:21-22), they co-participate in the postdeluge covenant (Gen 9:9-10), and they are both culpable for the taking of human life (see Gen 9:5; Ex 21:28). It will become clearer in the upcoming chapters that I cannot talk about humanity without also talking about the land, vegetation, animals, and the cultures that emerge, stratify, and reflect all of these relationships.[13]

The most difficult thing to understand here is that I will mostly be focused on demonstrating a conceptual analysis, comparing the ways these different groups of intellectuals thought about parallel concepts—paying particular attention to those things that the biblical authors shine their lights on. In this case, the parallel concepts are those that fund the narratives of Scripture and evolutionary thought about *Homo sapiens*: an explanation of scarcity, mis-fit to environs, and cultural mechanisms for sexual propagation.[14]

In speaking of nature, the definition of what is *natural* sits at the heart of the divergent visions of Israel's Scripture (in which I include the New Testament) and some in the evolutionary sciences. We must get clear on what we mean when our claims invoke nature.

NATURE AND THE SUPERNATURAL IN GENESIS

Our concept of nature today is both recent and complex. We can mean several things when talking about nature or describing something as natural. *Nature* can mean:

- the essence of something (e.g., the nature of truth)
- all the biochemical aspects that constitute earth
- an environment untainted by human intervention, also conceptualized as "the wild"
- without artificial components (where *artificial* means "made by humans")
- the web of life and all that supports it (i.e., the biosphere)

---

[13]Matthew J. Lynch, *Portraying Violence in the Hebrew Bible* (New York: Cambridge University Press, 2020), 1-69.

[14]*Homo sapiens* distinguishes what is commonly called "human" from other *Homo sapiens*, such as *Homo sapiens neanderthalensis*. Some would extend the term *human* to include any kind of *Homo sapiens*, and some label Neanderthals as *Homo neanderthalensis*.

More ideas about nature could be easily added to this list. When referring to the state of the cosmos, which nature we mean reveals what we think about both physics (crudely: how stuff works) and metaphysics (even more crudely: what stuff is and how it is related to other stuff).

The biologist's use of *nature* probably stems from the concepts given to us by nineteenth-century explorer Alexander von Humboldt. Humboldt might have been the first to suggest that *nature* ultimately refers to "earth as one great living organism where everything was connected."[15] Nature is the "web of life," as we have now come to think of it thanks to Humboldt. His conception of nature as interconnected climates and creatures created the ecosystematization of earth that Darwin later employed in his views of natural selection. Humboldt's version of nature is often assumed in discussions like these.

We can even talk today about the "laws of nature," which arose in the sciences as a strictly theistic metaphor. "There were laws for the whole of nature because there was a divine lawgiver."[16] That idea of divinely given laws of physics eventually drifted to a nontheistic version popular today. But we should notice that when we are trying to describe something about the implicit rules, forces, and functions of the universe, "rules," "forces," and "functions" act as political metaphors in a divine empire appropriated for physics. But for Darwin *natural* tends to mean something close to "biological processes unimpeded by human [or divine] intervention."

BIBLICAL NATURE VERSUS SUPERNATURAL

For a moment, I want to privilege the biblical authors' view of nature first (though they have no term for "nature"), which should help to explain why terms such as *supernatural* cannot help us to think about what is described in the biblical texts. In contrast to all these ways of thinking about nature, I will argue that we need to understand the biblical author's account of a theologically natural world.

No good terms exist, whether *un-*, *anti-*, *de-*, or *supernatural*. None of these fully capture what the biblical authors seem to be describing when

---

[15]Andrea Wulf, *The Invention of Nature: Alexander von Humboldt's New World* (New York: Vintage Books, 2020), 2.
[16]Jeffrey Koperski, "How the Laws of Nature Were Naturalised," *Science & Christian Belief* 33, no. 2 (October 2021): 63-82.

God acts in the cosmos. So, we will have to be good explorers ourselves and be satisfied by the Hebrew world of nature that they conceptualized. We must use the best metaphors and analogies we can find to illustrate their concepts of what we casually call nature today.

The biblical authors conceived of the cosmos as currently *unnatural* or, as some say, denatured. Here again, we have a clash of terminology with the various meanings of *natural* and their opposites: *natural/supernatural* or *natural/unnatural*. In the natural-unnatural opposition, biblical authors might argue that the Edenic world rightly oriented for flourishing is properly natural, while the disoriented bent that corrupts creation is properly unnatural. In coming chapters we will see how they seem to argue that this world east of Eden is fundamentally defunct: cancerous, decaying, warped, or any number of metaphors of malformation and internal disorientation.

The problem is endemic across creation and revealed in the aggregate as well as in individuals. Because the good and natural order no longer persists in its intended shape after Eden, majestic spots such as Yosemite's Half Dome and the Namib's sand dunes might be theologically understood as *unnatural*: malignantly warped even if functional. Our bodies, the dirt, our desires, and our efforts twist and contort with and against us. These all undergo a metaphysical disorientation even if the physical world is still theologically "very good." This is the state of the cosmos explored by the biblical literature and noticed by Darwin.

In its present state, our world is not the way it once was, which does not entail the absence of God's invisible powers. In the conceptual world of the authors, creation becomes twisted and cancer-ridden with respect to humanity even if sustained by God. More specifically, at the zenith of creation, at least one spot on earth—to blur Genesis 2 with Genesis 1—naturally constituted what God refers to as "very good" (Gen 1:31). In the garden at Eden, God showed us nature as intended, at least after the construction of woman (Gen 2:22). We will see this view of creation *denatured* in more detail in later chapters.

Regarding the natural/supernatural opposition, there are at least two difficulties with defining *natural* as opposed to *supernatural*. First, we often think of the supernatural world as invisible and thus mysterious to us. We might even think of the natural world as if it is the one visible to us and known by science, while the supernatural speaks to the invisible and divine causes behind reality.[17] However, the scientific enterprise has *not* traditionally focused on the visible world, whether discerned by the naked eye, electron microscope, or through spreadsheets full of statistical analyses. In reality, the scientist dedicates herself to the study of the invisible features of the visible world. An illustration might help here.

The chemist adds nothing to our understanding by describing what we can plainly see before us. The remarkable feat of the scientific community occurs when a scientist accredited by the community of chemists can tell us all the things that *we cannot see but that must be true* for the molecules *that we can see* to be the way they are.

The chemist's near-magical ability to "see," expound on, and successfully predict all the invisible features of this molecule make her the chemist and the rest of us the pedestrian learners. She can tell me why the Mickey Mouse shape of water molecules contributes to the surface tension of water, why the crystalline properties of ice are unique to water among other liquids, or how water joins to salt to slip more or less easily between biological cell walls. None of these things are visible in the molecule itself. All these invisible features are testable and predictable, to a point. The matching of her explanation with the data indicate her truer understanding of H_2O.

This is the problem: thinking of the supernatural as the invisible forces of our world simply does not help clarify how those "supernatural" functions are not part and parcel of nature as we experience it.

Second, the natural-supernatural divide cannot adequately accommodate the biblical literature's conception of the cosmos. That dichotomy tends to theologically normalize our present physical world, as if it were the way it is supposed to be. In this view of nature, the earth just is natural and good as is, even if corrupted, contorted, or whatever other metaphor captures creation's brokenness.

Nature for some would mean the untainted earth (or "the wild"), where human intervention is unnatural, akin to domestication. When reading biblical stories, this view might conceive of a God who supernaturally

---

[17]J. Richard Middleton's eloquent solution to this is to separate out "natural" from "cultural" conditions for human calling. Middleton, *A New Heaven and a New Earth: Reclaiming Biblical Eschatology* (Grand Rapids, MI: Baker, 2014), 42.

intervenes against the laws and the flow of nature to fix things in history, such as the founding of Israel by drawing them out of Egypt. Darwin gestures at this natural-versus-artificial divide in the prologue to his sixth edition to *The Origin of Species*. In making a contrast to the farmer's intentional selection, which happens in agriculture "by art" (i.e., artifice), Darwin notes that selection "seems to be done with equal efficacy, though more slowly, by *nature*."[18]

Natural, as an adjective, can also conjure a whole slew of associations in our minds that further complicate its meaning here: waterfalls, forests, predation, and even television shows such as *Mutual of Omaha's Wild Kingdom* all represent aspects of the natural world. *Natural* can strike us as the opposite of artificial or human-made. Thus, we can have a food product such as pasta that is both made by humans (artificial) and advertised as "all natural." (The recent switch to words such as *raw* now distinguishes what used to be marketed as "natural" in food sales.)

Natural can refer to the state to which things return when left alone by humans—a slice of pumpkin pie that goes back to its "natural state" in the compost heap (what a tragedy!). This idea of nature has a gravitational pull to it, bringing everything back down to a more basic state of existence. Nature can also be disrupted when humans intrude and build unnatural structures such as outhouses, span bridges, elementary schools, or investment banks.

Even more complicating, *nature* has been semidivinized into Mother Nature as a synonym for the collection of natural events neither created nor controlled by humans. In this sense, excess greenhouse gases are not natural, but the hurricanes resulting from the higher ocean temperatures are acts of nature—Mother Nature, even.

Thinking about Darwin's process of selection, *natural* describes the conditions that create change in organisms unimpeded or uncoerced by humans. But, for many of us today, *nature* has come to mean a world devoid of spirits and demons, absent any animistic gods or causal forces by persons or divine agents. Those spiritual realities belong to the "supernatural" realm, unseen by the human eye.

---

[18]Charles Darwin, *The Origin of Species, by Means of Natural Selection*, 6th ed. (London: John Murray, 1872), 7, emphasis added.

It is now obvious that we need another set of metaphors to help us think about what is and is not natural.

Hence, a sticking point between Darwin's concept of natural selection and the biblical account of selection might center on our conception of nature itself. For Christians, Genesis 4 and forward has been long read as an account of the world *unnatured* or nature that presently is not but could have been otherwise flourishing. Genesis seems to espouse that we actually and only have access to creation already warped by sin, even if it will someday be straightened and reoriented. In its present state, the cosmos is "not the way it's supposed to be," as one theologian elegantly simplifies it.[19] Or, as an ancient Jewish scholar once used more words to say:

> For the creation waits with eager longing for the revealing of the children of God, for the creation was subjected to futility, not of its own will, but by the will of the one who subjected it, in hope that the creation itself will be set free from its enslavement to decay and will obtain the freedom of the glory of the children of God. We know that the whole creation has been groaning together as it suffers together the pains of labor. . . . Now hope that is seen is not hope, for who hopes for what one already sees? But if we hope for what we do not see, we wait for it with patience. (Rom 8:19-22, 24-25 NRSVUE)

If they are correct, then biblical assessments of nature must be calibrated accordingly, not to the present or past conditions, but to another metaphysical situation to which the world was once oriented and will return in the future. This renewal requires divine creation and restoration, which puts it outside of the reign of entropy as it is traditionally construed.

For the biblical authors, we are stuck calling balls and strikes in a game with corrupt referees and players. It is akin to judging the brutalist concrete style of architecture by looking at an earthquake-rattled building. Some modest judgments could be made if the building merely has a fractured façade. But, how confident would we be in our aesthetic assessments if the building had been pulverized to dust with only the rebar-skeletal frame surviving? And which is its "natural" state—whole, cracked, or pulverized?

---

[19]Cornelius Plantinga Jr. brilliantly explores the pernicious and sophisticated ways in which the world as we know it is unnatural. See Plantinga, *Not the Way It's Supposed to Be: A Breviary of Sin* (Grand Rapids, MI: Eerdmans, 1996). See also Michael D. Williams, *Far as the Curse Is Found: The Covenant Story of Redemption* (Phillipsburg, NJ: P&R, 2005).

If we are tempted to say, "its brand new and whole state," then we now see the problem of beginning an account of selection with the presumption that the cosmos currently exists in its natural "whole" state, the way it was always meant to be.

Instead, all comparisons between the biblical account of origins and those of evolutionary science must acknowledge that the measurable world is metaphysically disoriented for the biblical authors. Yet, most of us fall into a habit of considering this warped universe properly *natural*. This is so much so that the term *nature* can also mean "the essential way in which something exists and always has." The biblical authors' conceptual story of the physical universe entails less stasis and more fracturing.

God's sustaining force and causative actions operate throughout the universe, even if the biblical authors do not portray it as God's direct action. Yet, for many of us today, scientific explanations about the invisible features of the physical world must be based in the orderly operations of so-called nature, with no divine causation behind the scenes. The earth, in particular, is free of God's actions, which means we can describe invisible features of reality such as osmosis, survival of the fittest, electromagnetism, and the rest without worries about divine influence. In this view, our understanding can be securely based on observations calculated strictly and generalized out beyond our ability to observe—no gods necessary.

The biblical authors also believed in basic lawlike operations of the universe but included God in the chain of causal events and considered that to be the way things are in a metaphysically disoriented universe. Biblical texts do not portray the flood, the plagues on Egypt, God's cleaving of the Sea of Reeds, and Jesus' quieting the Galilean Sea as supernatural intrusions into the natural world. Rather, the biblical authors depict decisive actions by God in history, who normally guides creation on a regulated course of affairs while allowing it to degrade under the curses of Eden or act askew from those regulations in signs and wonders. The precise ratio of God's sustaining acts compared to his permissive degradation of the cosmos is a topic for another book.

The point is: In comparing the conceptual world of evolutionary science with that of the biblical authors, we must be sensitive to the various conceptual schemes of nature. Failing to grasp how the biblical authors conceive of the natural state of the universe can cause us to lose the thread they are stringing for us. They presumed that at least one spot in the cosmos was as it was intended by God and then degraded in corruption after Eden. I would call that the theologically natural world. The present state of decay that we experience connects directly to a promised reorientation of the cosmos in its climax, a topic to which we will later return.

It should now be clear that what conception of *natural* the biblical authors intended determines how we understand the physics and metaphysics of their intellectual world. I will try not to import alien and modern notions of nature into their conceptual world, distorting our ability to make apples-to-apples comparisons. Such reasoning requires our imagination, even if through a glass darkly.

CHAPTER FIVE

SCARCITY ACROSS GENESIS

Of all the [parched hills] in all the [land bridges]
in all the world, [Israel] walks into mine.

RICK BLAINE, CASABLANCA (ADAPTED)

Behold, the eye of Yahweh is on those who fear him . . .
that he may *deliver their being from death*
and *keep them alive* in famine.

PSALM 33:18-19 (MODIFIED)

I have listened to the voice of Yahweh my God . . .
Look down from your holy habitation, from heaven,
and bless your people *Israel* and the dirt *that you have given us, . . .*
a land flowing with milk and honey.

DEUTERONOMY 26:14-15 NRSVUE (MODIFIED)

REMEMBER THOSE REALIZATIONS YOU had while peppering your mom or dad with questions about the wider world? I can picture now when I asked my dad about meteorology from the back seat of our "sunbrite yellow" Volkswagen Rabbit. I did not understand how adults knew whether we would have rain or shine tomorrow. The 1984 Tulsa Memorial Day flood had just drowned our city by surprise, killing fourteen people overnight with eight hours of rain. Heading downtown on Riverside Drive on the hot vinyl bench seat of that VW, I wanted my dad to explain how we knew what the weather would be like the next day.

He spelled out, as best he could to a ten-year-old, how meteorologists collect information about the temperature, barometric pressure, winds, sunshine, humidity, radar data, and so on. They put all that information into computer programs that make good guesses about tomorrow's weather. While I was focused on whether it was going to be hot or cold, rainy or not, meteorologists were tracking all kinds of data that I did not even know existed or mattered.

My dad introduced me to a whole other world of weather from the driver's seat of that VW Rabbit. I did not know that meteorologists think differently from everyone else about all the little things happening in our atmosphere. Their anxieties and excitements are sharply tuned into things most of us miss or ignore.

The same kind of mind-opening experience happened when my wife and I moved our family to Israel for my sabbatical. When we lived in Jerusalem and traveled all over Israel and the Palestinian territories, a whole other world opened up to me. Its soils, sun-brightened Judean rocks, and its cloudless and hot days taught me about the different anxieties and excitements that ancient Israelites would have been fretfully tuned in to. Jesus even used Galilean farmers' attentiveness to daily weather to chastise them. His diatribe on those farmers accuses them of missing the forest for the trees of God's coming kingdom. He calls them hypocrites for their overeager attention to the weather while ignoring the "present age" (Luke 12:54-56 my translation). Their anxieties about scorching winds and rain-pregnant clouds doubtless stemmed from their anxieties about scarcity, famine, and starvation.

Scarcity drives much of natural selection for Darwin. East of Eden, scarcity intermingles with violence in Scripture but also affords Israel opportunities to see God provide food despite its improbability. To understand the prevalence of scarcity in the biblical logic, we need to reorient ourselves to the on-the-ground realities of Iron Age Israel. Then we can see how Eden functions as an argument about Israel's trust in God himself, who is the antidote to scarcity and violence. Only then can we judge how the thinking embedded in Scripture's origins can contribute to concepts within theistic evolution—or not.

Fair warning: this first foray into the biblical thinking on scarcity will be the longest because it will set us up for later investigations into Scripture's views on fittedness and sex (which are shorter). Bear with me as we lay the biblical tracks we will follow in order to see the Bible's conceptual world emerge.

WE DO NOT KNOW EDEN

Ellen Davis begins her book on ancient Israelite agriculture by directly confronting what many of us do not understand: "Agrarianism is a way of thinking and ordering life in community that is based on the health of the land and of living creatures."[1] As a twenty-first-century megacity dweller, it took me a while to get my own thinking around just the first part of her sentence, "Agrarianism is a way of thinking," much less to grasp how community care, health care, and land care fit into a community's thinking.

For most of us in the so-called industrialized world, Genesis's powerful description of a garden at Eden has little, if any, purchase on us. The narrator of Genesis 2 describes a garden spot that might be nice as a day spa for you or me, but not much more. We might even hear the description of the garden and think, "Sounds nice enough, but *what's there even to do in Eden?*"

If we have lived in the West, even if we have lived in poverty in the West, we still have access to food, education, and many other resources unknown to all but medieval European kings. Our generalized concern about resources has atrophied, made flaccid by our birth-to-death opulence. Unwittingly anesthetized to the realities faced in the Majority World, we have been habituated into calorie-dense food delivered effortlessly to us by car in a currency-driven economy and consumed in climate-controlled homes. Hence, our bodies are numb and our ears deaf to what ancient Israelites would have heard in the telling of Eden.

Of today's 1.2 billion people living in some form of poverty, only a small subset of those people could even be described as having

---

[1] Ellen F. Davis, *Scripture, Culture, and Agriculture: An Agrarian Reading of the Bible* (New York: Cambridge University Press, 2009), 1.

experiential knowledge of the lived anxieties of Iron Age Israelites in Canaan.[2] At best, most of us can conjure up the significance of Eden by picturing an image of two nude adults in a lush tropical setting—a visual scene that only highlights our ignorance of ancient agriculture and geography in southwest Asia.

What about the Iron Age folks in the Bible? Israelites were spread across a wildly contrasting land. Some of them lived off quasi-arable soils in a land bridge between the warring empires of Egypt and Mesopotamia. Unlike those empires, Israel did not have abundant and ever-flowing rivers to sustain them.

In sobering ways, much of Canaan was the worst land in which God could have set his children. Israel's land was mostly riverless and perched on the largest *unusable* body of water in the region—the Dead Sea. Israel was dependent on diverting and trapping seasonal rains to survive. Crop failure was *the enemy* of the people when foreign invaders were not pressing in.

Yet, of all the parched hills in all the land bridges between empires in the world, God marched Israel into Canaan. Reading Genesis 1–11 from within the perspective of generational survival in Canaan, the theological description of the soil of Eden has more grip for them than it could for us today. Victor Matthews and Don Benjamin capture the life of most early Hebrew settlers who would have heard this account of Eden:

> These pioneers did not wage war. They survived war. . . . They were social survivors who fled the famine, plague, and war which brought the Bronze Age to an end. . . . They removed themselves from the urban centers of Canaan and settled into a politically less complex society. . . . The Hebrews were remarkably successful at maximizing their labor and spreading their risks.[3]

---

[2]Poverty here is defined by the Oxford Poverty and Human Development Initiative. There are "ten needs beyond 'the basics' in three broader categories: nutrition and child mortality under Health; years of schooling and school attendance under Education; and cooking fuel, sanitation, water, electricity, floor, and assets under Living Conditions. If a person is deprived of a third or more of the indicators, he or she would be considered poor under the MPI [Global Multidimensional Poverty Index]." See Tanya Basu, "How Many People in the World Are Actually Poor?," *The Atlantic*, June 19, 2014, www.theatlantic.com/business/archive/2014/06/weve-been-measuring -the-number-of-poor-people-in-the-world-wrong/373073/.

[3]Victor H. Matthews and Don C. Benjamin, *Social World of Ancient Israel 1250–587 BCE* (Peabody, MA: Hendrickson, 1993), 3-4.

"Spreading their risk" grew into the favored survival tactic in Canaan. Israel became a soil-, seed-, and water-shaped people. Their architecture, governance, taxes, townships, work ethic, borders, festivals, and more revolved around three factors: seasonal rains, return on diversified crop investment, and stable retention of the yield. Most simply, "Agriculture dominated Israelite daily life."[4]

Israel prayed for an appropriate amount of well-timed rain in order to survive. Rain out of season missed planting or harvesting. Too much rain caused floods that could wash away topsoil. Fields were carefully nurtured from arid to fertile over years. Israelites were some of the few who figured out how to improve their soil's biome through crop rotation and the use of fertilizers such as manure, compost, and ash.[5] Losing that cultivated soil meant years of lost investment, including its future returns. That loss could be disastrous.

Other lurking factors preyed on their livelihood, even if their watered soil came to harvest. Crop failure rates are estimated at 30 percent, which accounts for the Israelite diversification efforts that likely sought to "spread risk" of starvation.[6] Moreover, if a crop yielded, the ability to secure the food in storage over the year presented another major challenge to Hebrew hamlets. With only clay jars and stone silos, the looming threat of mold, rot, vermin, and theft confronted every homestead every year. If insects did not descend, mold and mildew subsided, and vermin did not attack, then the "milk and honey" of Canaan can be said to have flowed, as it were.

Now we might be able to envision what they saw in the spectacle of Eden. A safe and fertile garden-city must have seemed as implausibly utopian to the ancient Hebrew as it seems pleasant-if-uninteresting to most of us today. Ferdinand Deist thinks that the biblical garden texts reveal "humankind's picture of the ideal world."[7] Hence, that Eden-as-a-garden might not actually have a visceral appeal to us today signals that we are the ones estranged from the conditions that illuminate the text. "Against the

[4]Oded Borowski, *Agriculture in Ancient Israel* (Winona Lake, IN: Eisenbrauns, 1987), 5.
[5]Borowski, *Agriculture in Ancient Israel*, 145-48.
[6]Ferdinand E. Deist, *The Material Culture of the Bible: An Introduction* (Sheffield, UK: Sheffield Academic Press, 2000), 147; Matthews and Benjamin, *Social World of Ancient Israel*, 38.
[7]Werner Berg, "Israels Land, der Garten Gottes: Der Garten als Bild des Heiles in Alten Testament," *Biblische Zeitschrift* 32 (June 1988): 35-51, quoted in Deist, *Material Culture of the Bible*, 156.

background of the Palestinian climate and landscape, and the hard labor required for subsistence, it is quite understandable that a lusty garden would seem the ideal environment to an ancient Israelite." So, "[Eden] summarizes the dreams of an Israelite peasant: the dry soil is watered to produce a garden with perennial streams, full of beautiful, shady and fruit bearing trees, but no thorns or thistles. There is no hard labour but minerals galore, peace between humankind and animals, and the man with his wife are in charge of everything."[8]

Claus Westermann calls us to remember that the Hebrew creation myth (and *myth* does not mean "fiction" for him) can only be properly understood from an existentially threatened reading. Reading Genesis 1–2 from the threat of extinction might help us to understand: "*The myth belonged originally to the context of survival*, an expression therefore of one's understanding of existence, of one's understanding of the existence of the threatened-self. . . . Reflection on Creation meant to rehearse (i.e., to repeat by narrative), in the present world and in man's dangerous situation, the beginning, when what now is came to be."[9] The Iron Age reader would have seen the curse of the womb and dirt as earth shattering, quite literally—creating the conditions for mis-fit (which we consider in later chapters).

Just as I missed all the meteorological data that helps to predict weather, we might have missed the rhetoric of delight in the garden story. We nonsubsistence nonagrarians might also miss the striking terror of God's curses. The curse of the dirt creates not just inconvenience but the conditions for famine. In the logic of Genesis, famine can yield exploitation and violence apart from the normal fates of starvation, cannibalism, and death associated with it.

ABUNDANCE AND SCARCITY: GENESIS 1–11 AND BEYOND

The idea of scarcity enters the biblical story line at an unusual moment: in the description of a garden stocked with fruit trees "pleasing to the sight and good for food" (Gen 2:9). Prior to the in-the-ground story of humanity's creation,

---

[8]Deist, *Material Culture of the Bible*, 156.
[9]Claus Westermann, *Creation*, trans. John J. Scullion, SJ (Minneapolis, MN: Fortress, 1974), 12-13, emphasis added.

the cosmological creation of Genesis 1:1–2:4 already included provision of food: "Behold, I have given you every plant. . . . You shall have them for food" (Gen 1:29). In the garden, after fashioning "the dirtling" (the man) *from the dirt* and constructing the *woman from the man*, we find them coupled as "one flesh": naked (*'arummin*) and not ashamed (*lo yitboshashu* [Gen 2:25]).

In modern English, *naked* and *ashamed* appeal to something quite different from the biblical use of these terms. Across the Hebrew Bible, both terms suggest lack of safety and provision. While "naked" (*'arom*) means "exposed" in its most basic sense, biblical authors use it metaphorically for oppression (Is 58:7), physical vulnerability (Amos 2:16; Job 22:6), the threat of judgment (Hos 2:3), and most significantly lacking resources (see Job 1:21; Eccles 5:15) or without possessions.[10] *Naked* in most contexts signifies vulnerability, even helplessness because of scarcity or war.

Predictably, we probably read the term *shame* too strongly as well. In the garden, we see the united flesh of the dirtling and the woman exposed with sufficient food but without clothing. Yet, they bear no physical posture of shame. Our tendency to associate shame with an inner emotional state must be balanced with the outward focus that shame bears as well. As one Hebrew dictionary warns, "English stresses the inner attitude, the state of mind, while the Hebrew means 'to come to shame' and stresses the sense of public disgrace, a physical state."[11] It is a both/and situation, but our own connotations of *shame* probably tip the scales toward inner dispositions.

The couple is naked in that they are exposed, which is most often associated with persons facing famine, poverty, or defeat in war. They are not humiliated, which would be associated with a public failure or defeat. In the garden, "naked and not ashamed" paradoxically refers to both a current state of provision and a possible future state in the fields of scarcity east of Eden.

---

[10]As in stripped bare, lay bare of resources, or "without possessions." See "עָרוֹם," in *Enhanced Brown-Driver-Briggs Hebrew and English Lexicon*, by Francis Brown, Samuel Rolles Driver, and Charles Augustus Briggs (Oxford: Clarendon, 1977), 736.

[11]As in "put to shame" or "ashamed before one another." See "בּוּשׁ," in Brown, Driver, and Briggs, *Enhanced Brown-Driver-Briggs Hebrew and English Lexicon*, 101-2. See also R. Laird Harris et al., eds., *Theological Wordbook of the Old Testament* (Chicago: Moody, 1980), 2:50-59: "To fall into disgrace, normally through failure, either of self or of an object of trust." "English stresses the inner attitude, the state of mind, while the Hebrew means 'to come to shame' and stresses the sense of public disgrace, a physical state." In rare cases, it can mean "delayed" (e.g., Judg 5:28).

By connotations alone, we have all the hallmarks of the coming problem in Eden here at the end of this postscript about coupling (Gen 2:24-25). Yet, the couple exists unironically satisfied, a paradox reconciled in the text only by God's security and provision in the garden. God's provision, fruiting from their work, is captured in God's first command to the man alone—all stated in masculine singular: "Eating-ly you shall eat from all the trees in the garden" (Gen 2:16 my translation).

The garden functions as one location suitable for God's presence and a place where humans could be naked to the elements as fruitarians—yes, that is the correct term—made for each other and surrounded by desirable food. It lacks any noted violence or predation.

The garden in Eden is, quite simply, *an ancient agrarian subsistence farmer's paradise*, free from all the anxieties and pressures of securing and storing food, maintaining shelter, and defending against predators and illnesses, all while defenseless against empires. As many scholars note, God does not task them with sowing domesticated crops such as wheat or barley but eating the fruit produced by trees planted and tended for that task alone (Gen 2:9).

The garden's security becomes noticeable upon their breach. The couple is sent out eastward and is kept from returning by heavenly creatures with flashing swords.[12]

Immediately after their violation, three effects ensue: (1) there is a promised hostility between human progeny and the serpent's "seed" (Gen 3:15); (2) the relation of sexual coupling and the fruit of that coupling is fractured—the woman's desires for her husband and reproduction will now endanger her (Gen 3:16); and (3) the soil turns its productivity against humanity (Gen 3:17-19). In each effect, what did not have to be the case, in the original structures of Eden, now is the case. Each curse bends an otherwise amenable relationship to indicate a metaphysical change.

Before entering the first story of scarcity and violence, we can already anticipate some outcomes. If scarcity, insecurity, dysfunctional procreation, and dysfunctional desires curse the landscape of relationships east of Eden,

---

[12] As Joseph Blenkinsopp puts it, it is the tree of life itself being "guarded by the formidable *kĕrûbîm* and the flashing sword after their expulsion." Blenkinsopp, *Creation, Un-creation, Re-creation: A Discursive Commentary on Genesis 1-11* (New York: T&T Clark, 2011), 79.

then we should not be surprised to find efforts to buffer against these fraying complications by the characters of Genesis.

East of Eden, city building and entrenchment appear to mitigate the dirt's unyielding ways and protect against violence from scarcity. Technologies, such as the bronze plow, will force the cursed dirt to give up more of its strength than ever before, no matter how meager the spoils of that battle might be.

However, no flood, city, tower, or technology can tame unruly desires. Humanity's desire to entrench and overcome the curses through technology will be weighed and found wanting by God.

In the matters of scarcity and its connection to violence, we will discover in Genesis 4–11 a pattern of staving off scarcity through entrenchment, portable wealth, and the technology that commandeers security apart from God.

Cain: Slave of the dirt. For reasons known only offstage to the author, Cain works the cursed dirt. By shepherding, Abel frees himself from many associated problems of the cursed dirt. Cain, now labeled a "servant of the dirt" (Gen 4:2 my translation) cannot "work and keep" the orchard of Eden in order to grow food. Though many translations opt for "tiller of the soil," "servant/ slave of the dirt" better reflects that this particular phrase—*'oved 'adamah.* Though the man of Eden is set to "work" (*'avad*) and "tend" (*shamar*) the garden, the use of *worker* or *slave* changes in exile from Eden. Where *oved* occurs in later contexts, the "worker" is generally considered lowly.[13] Additionally, Noah will later be called a "man of the dirt" (*'ish ha'adamah*) as opposed to a servant (Gen 9:20 modified). The distinction seems clear: Cain does not merely work the dirt; the dirt works him! He slaves to reap its meager returns. Abarbanel, a medieval Jewish commentator, notes that in their vocational disparity, Abel "conducted himself like a free man, while Cain was a slave to the [cursed] soil."[14] Whatever the best translation—"tiller of the soil" or "slave of the dirt"—the narrative context makes clear that Cain's work is not identical to the man's working and keeping of the garden (see Gen 2:15; 4:2).

Cain brings an offering (*minkhah*) from this cursed dirt, but God shows no favor. Scholars ancient and modern have puzzled over the logic of God's

---

[13]Cf. Gen 3:23; 4:2 (*'oved 'adamah*); Isaiah's beasts of burden (Is 30:24); Zechariah's false prophets concealing themselves as slave boys (Zech 13:5).

[14]Michael Carasik, ed. and trans., *The Commentators' Bible: The Rubin JPS Miqra'ot Gedolot* (Philadelphia: University of Nebraska Press/Jewish Publication Society, 2018), 49.

disfavor toward Cain. The early Jewish commentary of the Mishnah supposes that Cain's gift itself was inferior. Hence, Rashi's commentary repeats one Mishnaic sage who supposed that Cain must have given flax (and everyone who has eaten flax said "amen!"). Kimhi, another medieval rabbi, suggests that Cain may have offered to God his leftovers or inferior grains.[15]

Hostility, now pregnant in human endeavors, quickly emerges in the eastward movement of the story toward Cain's and Abel's altars. The serpent, like Abel, evaporates from the story. We soon see hostility of the most nauseating kind, but not between the woman's and the serpent's seed, as promised. Indeed, Genesis does not portray a serpentine battle with Eve's offspring in its text, if at all. It leaves open the question whether it is the serpent's influence or "sin . . . crouching at the door" (Gen 4:7) that surreptitiously funds the violence that we see in Cain, or maybe it is both.

God's direct intervention to keep Cain's sin from pouncing on him might be surprising, specifically when it is ineffective. God's appeal assumes that violence is not the necessary outcome of his cursed work in relation to his brother. We might have expected a personal intervention by God to work with Cain, showing him the error in what he is planning. Yet, the next sentence in the story demonstrates that even a rational appeal by God is not enough to assuage whatever fuels Cain's murderous intent.

The story's structure suggests something amiss in Cain's desires, where God equates his fallen face with his ruling over his desire to do violence. The language of Cain's ability to rule over his desire parallels the woman's curse:

> Your desire will be for your husband, and *it* will rule over you. (Gen 3:16 my translation)[16]
>
> Its desire is for you, but you must rule over it. (Gen 4:7 NRSVUE modified)

---

[15]Carasik, *Commentators' Bible*, 48.

[16]This independent pronoun (*hu'*) could be translated as saying that "he" or "it" (i.e., the longing) will rule over her, not necessarily her husband. The only other use of this phrasing with desires and ruling appears in the very next story with Cain. The parallel language and structure of God's argument to Cain—"its desire is for you, but you must master it" (Gen 4:7)—and the sequence of ensuing narratives reenacting the taking, giving, and listening to the voice (here that of Eve, then of other matriarchs with desires awry in Genesis), suggest that a reasonable translation of the woman's desires would be "your desire will be for your husband and it will rule over you." See Gen 3:6; 16:2-3; 27:8, 13, 15-17, in contrast to Joseph's refusal to "listen" to Potiphar's wife's sexual advance day after day, he notes that he cannot lie with her using the repetition of "all the things" (Gen 39:4-6, 8-9) that Potiphar has "given" under his

Notably, the cursed soil to which Cain has slaved soon becomes a blood-soaked vessel of his violence. The text suggests a link between Cain's cursed soil of scarcity and the violence he will commit. But God's intervention suggests that the connection is not necessary.

Several clues in the story point to an intimate relation between the soil's scarcity and what happens next. Cain hastily dispatches his brother's blood back into the dirt from which their father came—the same dirt from which his brother's blood will cry out. As Matthew Lynch argues, the spilling of Abel's innocent blood creates a primary "grammar of violence" across the thinking of the biblical authors.[17] Soil and violence do not mix.

Upon Abel's murder, the story repeats elements of Genesis 3. God comes to Cain seeking Abel with the question, "Where is Abel?" Excuses ensue. Then, God pronounces a curse similar to Genesis 3 on the dirt's fecundity.

Cain's curse exiles him from the previously cursed dirt that has "receive[d] your brother's blood from your hand" (Gen 4:11). This curse doubly endangers Cain, which he immediately realizes. He goes from an already cursed subsistence to instability caused by wandering. God curates Cain's punishment to him, an ineffective punishment if, say, the tables were turned and Abel had murdered Cain. Wandering cuts a "slave of the dirt" off from his means of production in a way that would not affect a nomadic occupation such as shepherding. The punishment suits the criminal more than the crime. Then, God effectively commutes Cain's capital sentence, as his crime was one that will later require execution by the community (Gen 9:5-6).

Notice that the ground will not yield to Cain, which forces him to wander. His murder of Abel will start a cycle of killing that requires divine interventions of biblical proportion. Cain's descendants emerge as the technologists (Gen 4:17, 20-23), settling and overcoming the

charge, "except yourself, because you are his wife" (Gen 39:4, 9). Joseph's intensive all-things-given-except-for-one-thing rhetoric echoes the Eden narrative strongly, forcing us to consider it in that context. See Robert I. Vasholz, "'He (?) Will Rule over You': A Thought on Genesis 3:16," *Presbyterion* 20, no. 1 (1994): 51-52; Dru Johnson, "The שמע–קול Motif," in *Epistemology and Biblical Theology: From the Pentateuch to Mark's Gospel*, Interdisciplinary Perspectives on Biblical Criticism 4 (New York: Routledge, 2016), 41-48.

[17] Matthew J. Lynch, *Portraying Violence in the Hebrew Bible* (New York: Cambridge University Press, 2020), 30.

cursed dirt problem. Eventually, Cain's offspring commit violence reminiscent of the murderer's forefather, this time as a rash killing of a young
man (Gen 4:23)[18]—later deemed a capital crime in some but not all cases
in the Torah.[19]

Violence is not always tied to scarcity. Genesis portrays the two in tandem
east of Eden, inaugurated by scarcity-inducing curses against the dirt. The
final fruitfulness of the man and woman east of Eden adds the third sexual
punctuation to Genesis 4 (notice the formula "and X knew Y and she
bore Z"; see Gen 4:1, 17, 25). Seth (*shet*) is a replacement, so named because
of the family's fratricide, "God has *appointed* [*shat*] for me another offspring
instead of Abel, for Cain killed him" (Gen 4:25). In violence, Adam and Eve
lost their son. In sexual reproduction, they gained the possibility of progeny
outside Cain's line.

Genesis does not portray violence as necessarily derivative from scarcity
in the initial stages of human history. This new and warped world metaphysically wrecked by the dirtling and his wife anticipates such violence, and
so does God. In fact, God's failed intervention with Cain teases apart violence from scarcity before the first violent act ever occurs. God reasoned
with Cain that his violent desires did not have to actualize.

Whether the reader presumed violence was necessarily connected to
barren dirt or not, God connects the violence and scarcity before and after
the first murder. But for God, violence does not have to result from scarcity.
That might be what God was warning Cain to do: separate his resource
scarcity from his plans to do violence.

Noah's cursed dirt problem. Why does God send the flood? The answer
appears straightforwardly in the text: "the wickedness of man was great in
the land, and every intention of the thoughts of his heart was only evil

---

[18]Middleton cites Lamech's bigamy in Gen 4 paired with his "revenge killing" (Gen 4:23) as in-
troducing the topic of violence against women. I agree in principle, and knowing his work, he
has more behind this claim than an opinion. However, ancient polygyny was sometimes used
to protect vulnerable women. Hence, it is safest to restrict ourselves to the overt and stated violence in the passage. J. Richard Middleton, *A New Heaven and a New Earth: Reclaiming Biblical
Eschatology* (Grand Rapids, MI: Baker Academic, 2014),

[19]Jonathan Burnside argues that violence such as Moses' rash murder of the Egyptian was of a
permissible sort—a "heat of the moment" killing. See Burnside, "Exodus and Asylum:
Uncovering the Relationship Between Biblical Law and Narrative," *Journal for the Study of the
Old Testament* 34, no. 3 (2010): 243-66. He also considers the horns of the altar and the need for
priestly remediation in all cases of asylum.

continually" (Gen 6:5 modified).[20] Yet, after the flood, the problem of humanity's evil clearly continues: "for the intention of man's heart is evil from his youth" (Gen 8:21). In other words, humanity's evil does not act as the central conflict of the flood narrative because the logic of narratives demands that a central conflict be resolved.[21] Humanity's evil hearts do not get fixed.

What is the story's conflict? It appears to be tucked away in the genealogy of Noah, rooted in the meaning of his name. Noah's name (*noakh*) plays on his commission "to relieve" (*nakham*) the curse of Eden: "Out of the dirt that Yahweh has cursed, this one shall bring us rest [*nakham*] from our work and from the painful toil of our hands" (Gen 5:29 modified). Indeed, this conflict is resolved and stated, in contrast to the problem of humanity's evil (Gen 8:21).[22]

What resolves the central conflict of the flood narrative? Productive dirt and the expansion of humanity's diet. "I will never again curse the dirt because of man," despite the fact that the "intention of man's heart is evil from his youth" (Gen 8:21 modified). Note that the ensuing postscript focuses on controlling for scarcity: "seedtime and harvest . . . shall not cease" (Gen 8:22). Human violence is linked to the wickedness of humanity, and it will rear its head soon again in Genesis. But this story's conflict comes all the way back to the garden's curse of the dirt.

In the same breath of divine pronouncements about the cursed dirt, murder is now said to be a capital crime for beasts and humans alike. Despite a more food-rich environment opened up to the humans, violence is assumed as ever-present. God's Noahic treaty anticipates animal and human violence, which will be reckoned from both animals and humans (Gen 9:1-6).

---

[20]Or equally, because the land (*'erets*) was corrupt (*shakhat*, like death's decay, possibly wordplay with *nakhat*, from *nuakh*), the land (*'erets*) filled with violence (Gen 6:11).

[21]Though I am describing here the narrative framing of the flood story (Gen 5:28–9:16), I would also assume that something like Wenham's chiastic formal structure (Gen 6:10–9:19) interweaves more discretely with this indelicate narratival structure as well. Gordon Wenham, "The Coherence of the Flood Narrative," *Vetus Testamentum* 28 (1978): 336-48.

[22]Noah's name already draws us back to the garden, the only other place where we see this term used up to now in Genesis (in a different grammatical form). God "rests/settles" (*nuakh*) the newly made man into the newly planted garden to work it and keep it (Gen 2:15). The biblical author reserves the only other use of this term translated "rest/settle" (*nuakh*) in Gen 1–11 to describe the ark when it "settled/rested on the mountains of Ararat" (Gen 8:4). Unsurprisingly, the ensuing language of the Noahic covenant hails from the creation of animals and commissioning of humanity in Gen 1. The wordplay appears intentional if not systematic when compared to the rest of Genesis, tying creation, fratricide, and flood into a distinct literary relationship.

Scarcity does not necessitate violence. And by making the dirt fruitful again despite the evil of humanity's hearts, God no longer maintains that violence has scarcity to blame, further emphasizing the contingency of their linkage.[23]

The Noahic covenant also constrains God's violence. God promises no more divine violence against all flesh, which refers to all living things with breath of life (*nephesh khayot*; see Gen 2:7, 19).

After the flood and God's bow-shaped treaty with animals and humans in parallel, Noah becomes a "man of the dirt." He builds a vineyard—mirroring the garden's agriculture, he does not plant crops—and then becomes inappropriately "naked" (*'erwah* in Gen 9:22, similar to *'arom* in Gen 2:25) from his drunkenness. The fruitfulness of Noah's soil exposes his foolishness and provides an opportunity for Ham to commit some kind of sexually tinged transgression against his father.

Scholars have noticed the parallels between the garden man and Noah. However, the focus on the curse of the dirt and its relation to scarcity has attracted less attention.[24] The flood appears as a restart, with the soil metaphysically returned to a more amenable relationship with humanity. The text seems to say: despite the soil's and the woman's fragile fertility, Noah and family should keep the commission to be fruitful, multiply, and spread out in the land. Noah's folly will not be the last time we see a drunk father's sexual encounter with his children (Gen 19:31-36). But it does end in pronounced curses against Ham/Canaan (Gen 9:24-27) and the eastward movement of Ham's descendants into the plains of Shinar in order to build a city called Babel (see Gen 10:10; 11:2).

THE DIRTLING	NOAH
Made from dirt	Called to survive flood for the sake of the dirt
"Settled" into garden	"Settled" into Ararat mountains
Fruitful marriage and food	Fruitful marriage and food
Inappropriate consumption of fruit	Relief from cursed dirt
Cursed dirt, death as return to dirt/dust	Inappropriate consumption of fruit (not grains)
Exiled eastward	Cursed Ham/Canaan, who settled eastward in Shinar

Figure 5.1. Eden and the flood compared

---

[23]Thanks to Celina Durgin for noting this connection to me.
[24]Blenkinsopp, *Creation, Un-creation, Re-creation*, 79.

The flood account homes in on humankind's relation to barren dirt and reveals divine justice that distinguishes violence as a nonnecessary result from scarcity. In other words, the violence that generates God's diluvian response depends on understanding that it—humanity's violence—did not have to be humankind's response to scarcity. God's postdiluvian response features a treaty with animals and humans to restrain his violence. This suggests that there was some kind of cooperative way imaginable, even if it was not followed.

The flood was not God's response to humanity's forlorn toil but to their ubiquitous savagery in the face of paucity. As shown with Cain, humanity's violence, which "ruined" (*shakhat*, Gen 6:11-12) the land, was unnecessary.[25] This means that the flood was avoidable. We cannot fail to miss this basic point because it features centrally in the story of Israel. Israel will be tested annually with divinely orchestrated scarcity in Sinai and later in Canaan. Now we continue eastward with Ham's descendants and the story line.

Babel's fertility and security. When we think of the biblical Babel, we think of a tower or maybe a city. But we should probably picture fields of food instead. "It has been known from antiquity that one of the principal advantages of the Mesopotamian area was the region's great agricultural potential."[26] James Breasted—who coined the phrase "Fertile Crescent"—sums up the fecundity of the Plain of Shinar: "When properly irrigated the Plain of Shinar is prodigiously fertile, and the chief source of wealth in ancient Shinar was agriculture."[27] Lying between the two great rivers, Euphrates and Tigris, and equivalent in size to the US state of New Jersey, Shinar was known for its agri-politics.

The abundance of strong, regularly flowing rivers and adequate fertile soil for a seed's purchase meant that agriculture also shaped how Mesopotamia thought. Many scholars believe that field and flock fertility in the region directly influenced the formation of its political structures. Even the renowned King Hammurabi was intimately concerned with the "trivial details" of the state's agricultural production.[28]

---

[25]Lynch, *Portraying Violence in the Hebrew Bible*, 54-56.
[26]Maria deJ. Ellis, *Agriculture and the State in Ancient Mesopotamia: An Introduction to Problems of Land Tenure* (Philadelphia: Babylonian Fund, 1976), 1.
[27]James Henry Breasted, *Survey of the Ancient World* (New York: Genn, 1919), 59.
[28]Ellis, *Agriculture and the State*, 5.

But for most Israelites, the water and fertility situation was precisely the opposite: "*Abundant and reliable water is precisely the commodity that the Israelite farmer lacked.* In this, he differed from his ancient Near Eastern neighbors in a way that made itself felt in aspects of his culture and lifestyle, among them his drink. This fact set him apart from the farmers of Egypt and Mesopotamia."[29] Some have argued that emergence of Babylon's dynastic state correlated directly with the creation of the irrigation canals of the region.[30] Water shaped their world, too, even their political realms.

In other words, Ham's descendants did not choose to settle in Shinar at random. Later in Genesis, Abraham's nephew Lot is also noted for choosing a morally impoverished location—Sodom—because of the land's tempting fertility: "Lot lifted up his eyes and saw that the Jordan Valley was well watered everywhere like the garden of Yahweh" (Gen 13:10 modified).

Ham's descendants build "a tower" (*migdal*) in Babel, a term that is mostly used to describe a military defensive position or a watchtower over a field or vineyard (see Gen 11:4-5; Judg 9:49-51; 2 Kings 9:17; Is 5:2).[31] This tower presumes that some kind of resource scarcity requires protection of plenty. Much the same way that today's militaries with forward operating bases in hostile territory usually build weapons towers for defensive reasons, we find towers in the Hebrew Bible used to thwart the violence of enemies or unwelcomed thieves of the human and animal varieties. In the flat plains of Shinar, a high position is needed to anticipate attacks and defend. Is the "tower of Babel" a definitively militaristic structure for defending crops? The story leaves the matter unclear. However, the connotations of such a fortification are unmissable in the Hebrew Bible.

Of course, the biblical story of Babel does not focus on abundance and scarcity. However, it would seem implausible that the agricultural potential of the land of Shinar was not one of the people's primary reasons for settling there.

But the conflict of the Babel narrative does not center on the tower either. The tower functions as a mere symptom of the problem, which creates a

---

[29]Carey Ellen Walsh, *The Fruit of the Vine: Viticulture in Ancient Israel*, Harvard Semitic Monographs 60 (Winona Lake, IN: Eisenbrauns, 2000), 27, emphasis added.

[30]Ellis, *Agriculture and the State*, 2.

[31]Thanks to Carmen Imes (and Richard Middleton) for reminding me of this point. J. Richard Middleton, *The Liberating Image: The* Imago Dei *in Genesis 1* (Grand Rapids, MI: Brazos, 2005), 185-234.

wordplay connecting Babel's foolishness: making a great "name" (*shem*, Gen 11:4), Shem's genealogy (Gen 11:1-32), and God's promise to make Abram's "name" (*shem*) great (Gen 12:2). Neither can the problem at Babel be the unity of language among these particular descendants of Noah (more in chap. 9).

The Babelians themselves utter the conflict of the story. "Let us build ourselves a city . . . lest we be dispersed over the face of the whole land" (Gen 11:4 modified). The resolution to the story's conflict is not the tearing down of the tower but the scattering of Babelians, stated twice: "Yahweh scattered them over the face of all the land" (Gen 11:8-9 modified).

Their offense stems from their urbanized rebellion against God's commission to fill the land in the creation of humanity (Gen 1:28), restated to Noah (Gen 9:1). God's solution scatters them away from the safe and fertile land they know. Like the tower that was instrumental for Babel to "make a name [*shem*]" for themselves, God's scattering in order to create new languages appears to be instrumental. Divergent languages ensure that the scattering takes hold.[32]

But why do these people hunker down in Shinar? The most obvious answer to the question: the dirt's fertility and abundance of water. The author comments directly on their ability to build a city in connection with the ability to make artificial stone (i.e., "they had brick for stone," Gen 11:3) and a natural source of mortar (i.e., bitumen), both regionally abundant. Shinar offers security in their relation to the fecund dirt and foreigners—both of which are regulated in Israel's coming instruction in the Torah (e.g., Lev 19:9, 33-34). They can in effect industrialize their shelter and agriculture. Joseph Blenkinsopp notices that their unity of language functions as part of the political project of Babel: "The city and tower are then revealed as a means of concentrating political power legitimated by potent religious symbols, and the settlers' resistance to linguistic differentiation is seen to be dictated by an awareness of language as an instrument of power, control and coercion, which it certainly is."[33]

How would Iron Age Israelites have heard the story of Babel? Jacques Ellul thinks that in Genesis, cities rise in the era of sinful humankind, when

---

[32]After writing this, I discovered that Ellul makes this same point about the instrumental and noncentral role of the tower and languages in the Babel story. See Jacques Ellul, *The Meaning of the City* (Grand Rapids, MI: Eerdmans, 1970), 15.

[33]Blenkinsopp, *Creation, Un-creation, Re-creation*, 79.

humans attempt to gain security and regain what was lost in Eden without looking to God. But he goes even further to connect Nimrod's ancestors, which are all named city-states of repute (Gen 10:9-10), with organized state violence. "The City is now a center from which war is waged. Urban civilization is warring civilization."[34] Citing Shinar's reputation across Scripture—a warring faction (Gen 14) and an object of Achan's temptation (Josh 7:21)—Shinar represents the opposite of peace.

Divine punishment for Cain and now Babel concentrates on removing physical and food security. The scattering of Babel fundamentally returns them to lack of security and renews their anxieties about scarcity. Languages and towers are prominent features but act only instrumentally in this story toward God's original commission of humanity. I will contend in chapter nine that the diversity of languages is fully and ordinarily explained in the genealogies of Noah (Gen 10). In the end, Babel is a narrative about an organized human attempt to stave off scarcity and physical threat in light of a divine blessing to fill the land.

Conclusions to Genesis 1–11 on scarcity. Throughout Genesis 1–11, the biblical packaging of scarcity and violence appears to offer a tentative critique of a necessary connection between the two. The curse of the soil creates hardships that devolve into violence (i.e., Cain, Lamech, and the days of Noah) or vain stabs at procuring one's own security, possibly even through technologies (e.g., Jabal, Babel). Unfortunately, various scarcities are the garden in which the biblical characters' insecurities grow.

Before moving into scarcity among the patriarchs and beyond, let me state succinctly what I take to be relevant so far to highlight the contrast with the patriarchal stories of scarcity:

- God creates a cosmos and a garden in Eden without violence, unlike most ancient Near Eastern creation narratives.

- Eden is secure (i.e., no fear of being stripped naked of resources or of being publicly shamed in defeat) and provides food when worked by humanity.

- Humanity is formed and "settled" (*yanakh*) into the garden to work and keep it up.

---

[34]Ellul, *Meaning of the City*, 168.

- Humans are coupled by their reproductive biological differences in Eden.

- Upon violating the singular instruction regarding their abundant food, curses ensue that include promised skirmishes, deteriorated bodies, the reign of desires, and scarcity sown into the cursed, misoriented dirt.

- Scarcity occasions the violence of Cain, which conjures more scarcity for a wandering Cain and rash violence in his descendant Lamech.

- Violence has no necessary connection to scarcity. Violence is depicted as a reasoned choice (i.e., God attempted to reason with Cain) based in unruled desires (see Gen 3:16; 4:7).

- Technologies are coupled with attempts to escape scarcity and violence (e.g., bronze and iron, bricked and gardened cities, tents and livestock).

- Violence and cursed-dirt scarcity continue to be coupled thematically, leading to the flood.

- Postflood, God releases creation from utter scarcity (i.e., a profoundly cursed dirt) through an omnivorous diet. The flood, however, does not entirely thwart humanity's evil and violence.

- Attempts at urban security and industrialized agriculture through technology are broken up by God through scattering: causing a return to the anxieties of scarcity and vulnerability to violence.

The patriarchs: Renewal of abundance. From the call of Abram forward, the favor of God comes wrapped in abundance of food and security from violence. God promises such security through treaty with Abraham and keeps his promises as Abraham fathers children and endangers his family in various episodes.

Like Abel, Abram is a shepherd, which buffers scarcity through milk, and meat when necessary, in the nomadic form of sheep, cattle, and goats. Still, famines force him to migrate, and violence is once again coupled with scarcity. Migrating into Egypt, Abram anticipates that his wife's beauty will endanger his own life and reacts by deceit. Without moral commentary from the narrator regarding Abram's solution to violence, God intercedes on behalf of Sarai's sexualized body with plagues that likely aim at creating the fear of scarcity (Gen 12:17). This sends Abram back out wealthier than when he entered Egypt (Gen 12:20).

Abram's livestock continues to safeguard against scarcity in Canaan (Gen 12:16; 13:2; 14:16, 21-22; 20:14), even when he repeats his deceit and hands over his wife's body to be sexually used by a foreign king (e.g., Gen 20).

In Genesis and beyond, the biblical authors maintain a distinction between scarcity and violence. Lot seeks security by choosing the land of abundance for himself (Gen 13:10). Yet, he settles and keeps his family in a city he knows to be profoundly violent (Gen 19:4). In fact, this exact scene of community-saturated violence will eventually be retold, at times word for word, in the book of Judges (see Gen 19:1-11; Judg 19:14-28). But, in the re-enactment, it is the Israelites themselves, the tribe of Benjamin, who descend violently on a Levite. The narrator portrays the Levite as foolish, gluttonous, and icy-hearted, tossing his wife out to be raped by a mob and then greeting her dead body with, "Get up, let's go" the next morning.

Violence has now been fully detached from scarcity, and Lot's story becomes Israel's story in the book of Judges (Judg 19:16-30), the lowest form of communal degradation that biblical authors can paint for us. At first tangled together, scarcity and violence now appear as two separate and distinct trajectories. Though Sodom sat alongside the well-watered Jordan Valley—"the garden of Yahweh" (Gen 13:10 modified)—abundance is not enough to counter Sodom's cultivated savagery.

Hopping back to Abraham, an odd story line about scarcity will emerge post-Sodom. Though they wander like Bedouins, who famously extract sustenance from the deserts, Abraham's descendants will all be blessed with abundance. Isaac gains bounty by repeating Abraham's pattern, handing over his wife for sex with a foreign king out of fear of violence (Gen 26:7). Jacob secures his wealth by deceiving his dying father and then deceiving his father-in-law (see Gen 27:18-29; 30:25-42).

Though they dwell in the same land, Abraham and his children traverse the land as shepherds, not entirely dependent on the seasonal rains like the later Israelites who settled Canaan. Water is disputed in Abraham's time according to well rights in the Negev desert, but Abraham seems unaffected by the utter dependence and covenantal testing that seasonal rain will later create for Israel. Hence, we should expect matters of water and abundance to be handled differently for "a wandering Aramaean" as opposed to the urbanized Iron Age Israelites later settled in Canaan.

Finally, in Joseph, we find a character in the story who breaks the arc. For this reason, Stephen B. Chapman suggests Joseph might be best understood as a type of anti-Adam (Gen 41:46-49). The Joseph narratives appear to connect more than superficially to the garden story. Joseph is the first person in the book of Genesis who does not listen to a woman with askew desires (see Gen 3:17; 16:2; 27:8-13; 39:10). Through his obedience to God, Joseph accepts violence and injustices done to him and faithfully arranges the work of the dirt to produce food that will save Egypt in a time of famine.[35] By saving Egypt from the famine, Joseph also saves his own family—though that was not his goal. Joseph has his own struggles with exploiting people, but his ability to exploit (later condemned in the Hebrew Bible) hinges on food scarcity (Gen 47:13-26). Genesis wants us to imagine that his exploitation of the Egyptians was not necessary.

In sum, all the patriarchs and their children prosper, resource-wise, according to God's treaty with them. Despite whether they act morally, courageously, or trustingly, and apart from violence in times of scarcity, the patriarchs gain an abundance of food while female fertility in their clan suffers (more on fertility in chap. 11). Unlike the people of Israel to come, abundance flows from the Abrahamic treaty, though not exclusively through Abraham, Isaac, or Jacob's fidelity to God's treaty with them. In the exodus from Egypt and forward, abundance will be distinctly tied to treaty obedience. Throughout all these texts, violence gets parsed out as a separable reaction to scarcity and abundance alike, not a necessary entailment of scarcity.

---

[35]Stephen B. Chapman, "Food, Famine, and the Nations: A Canonical Approach to Genesis," in *Genesis and Christian Theology*, ed. Nathan MacDonald et al. (Grand Rapids, MI: Eerdmans, 2014), 323-33.

CHAPTER SIX

SCARCITY BEYOND GENESIS

IS SCARCITY A TOPIC OF CONVERSATION in the exodus or Israel's entrance into Canaan? We have seen how Genesis lays out a fundamental problem with creation after Eden: scarcity due to the barren dirt, which creates conditions ripe for violence. The flood account partially corrects the dirt's hypersterility, but without correcting humankind's ever-evil minds (Gen 8:21). We must continue to follow the remarkable inclusion of agriculture interwoven into Israel's national story of justice and righteousness.

In Exodus, God tears down Egypt's industrial agricultural complex, yet God brings Israel into barren wilderness. And while we might think that the "land of milk and honey" offers the cure-all to the prolonged fears of scarcity, it might not mean what we think it means. Even the eschatological images given by the prophets of the Old and New Testaments recast the end goal of creation in light of our fears about want.

Ultimately, we must weigh the ever-present anxieties of scarcity in Israel that would have underwritten Scripture's larger story. Once we see what the biblical authors are doing with scarcity, we can think about how they might agree with or critique the insights of Darwin and later evolutionary scientists. They share a vision of violence that can erupt but also view such competitive violence as nonnecessary. While evolutionary science has favored cooperative models of evolution more recently, the biblical authors focus on cooperative trust under divine provision. As we will see, the "invisible hand" of Israel's subsistence economy is God's personal care for the land. The similarities and divergences require a closer look at how the biblical authors develop scarcity and violence beyond Genesis.

Because we are tracking the biblical concept of scarcity over many texts, we will need some endurance for the race. Hang in there as I work through the topic through many different biblical texts and angles. The hope is that a pattern of discourse about scarcity will emerge across these texts. I begin back in Egypt after the death of Joseph.

THE EXODUS

What happens to the topic of scarcity in the story of Exodus? Austerity drives the plot to bring an "anticreational" Pharaoh to his knees. So says Terence Fretheim:

> The culprit this time is not a serpent or a brother-killing Cain or the sons of God but "a new king over Egypt." . . . The focus is thus placed on him, not simply as a historical figure, but as a symbol for the anticreation forces of death which take on the God of life. The narrator's concern is with this king's response to God's extraordinary creative activity. This is a life-and-death struggle in which the future of the creation is at stake.[1]

The life-and-death struggle between God and his adversaries persists through the exodus, into the conquest and settlement of Canaan, and beyond. Because scarcity and violence eventually become the antithesis of prophetic visions of the eschaton, the pairing has secured its role as nemesis in the drama across both Testaments.

The exodus, the wandering, and Canaan: Scarcity as a test. Exodus redeploys creation's theme of fruitful generation and filling the land in its introduction: "But the Israelites were fruitful and prolific; they multiplied and grew exceedingly strong, so that the land was filled with them" (Ex 1:7 NRSVUE). Again, their fruitfulness creates a conflict in the story. The first Pharaoh of Exodus sees their multiplication neither as an ordinary good nor as resulting in a group of future Joseph-esque saviors of Egypt. Rather, they are viewed as a potentially violent threat to Egypt's security (Ex 1:10).

As the exodus story opens, we notice that war and insurrection are Pharaoh's chief anxieties, but also the presumption that war creates scarcity (Ex 1:11). Corresponding to this Pharaoh's anxieties, Yahweh targets the next Pharaoh's industrialized agricultural empire with the violence of his plagues.

---

[1]Terence E. Fretheim, *Exodus*, Interpretation (Louisville, KY: John Knox, 1991), 27.

Notice that most every plague of the ten targets Egypt's agricultural power and not its military strength—though they are inextricable from each other:

- blood (presumes a fish kill, wrecks irrigation and drinking water; Ex 7:19)
- frogs (frog kill in the fields, polluting crops; Ex 8:13)
- lice (afflicts "man and beast" alike; Ex 8:18)
- flies ("land was ruined by the swarms"; Ex 8:24)
- livestock death (Ex 9:6)
- boils (afflicts "man and beast" alike; Ex 9:10)
- hail ("on man and beast and every plant of the field" and servants of the Egyptians; Ex 9:22)
- locusts ("eat what is left to you after the hail"; Ex 10:5)

Only the final two plagues turn their sights from Egypt's agriculture. The ninth plague seems to degrade its god of the sun, Amun-Re, with darkness. The tenth plague then strikes reciprocally at the Egyptian boys as the first Pharaoh of Exodus struck at the Israelite boys.

Among other reasons, the stated focus of each plague indicates that they were meant to strip a well-watered and agriculturally sustained empire down to bare survival. Yahweh also has similar plans for Israel by leading them into the waterless wilderness. The way up was down, through dependence on God to provide food and water in times and locales of scarcity.

In the Sinai wilderness, scarcity was not punishment, as it was for Cain, but a test of fealty: Will Israel follow God's instruction through Moses? In the initial days after crossing the Yam Suf (Red Sea), dehydration focuses the Hebrews' complaints to Moses. God instructs Moses to make bitter water potable by throwing a log into it (Ex 15:25). Dehydration averted.

God then "tested" (*nasa*) Israel with lack of food and water. These tests are framed with the first intensive use of the well-worn phrase in the Torah: "*listening-ly listen to the voice* of Yahweh your God" (Ex 15:26 my translation). The emphasis on scarcity for the sake of testing is to see whether Israel will listen to Moses.

Scarcity acts as a means of revealing Israel's true loyalties. While this might sound petty on its surface, Israel's trust in Yahweh contrasts with their fantastical memories of Egyptian food. Immediately after the water is made drinkable, the romantics among them whine and recall "when we sat by the meat pots and ate bread to the full" (Ex 16:3). Though this memory is difficult to square with Exodus's account of their "bitter" slavery, God promises bread "that I may test them, whether they will walk in my instruction or not" (Ex 16:4 modified).

So, God sends manna for them to eat each "as much as he can eat," with only one Sabbath-reverent condition (Ex 16:16-19). The echoes of the garden at Eden swell to the point of unmissable here: "eating-ly eat from every tree," but with one exception (Gen 2:16-17 my translation). Fretheim reminds us that this is no mere entrapment but has a goal of imbuing Israel with knowledge (see Ex 16:4, 12): "The idealized and unwarranted memories of Pharaoh's food (v. 3) are to be replaced with the genuine memories of the bread from God (vv. 32-34)."[2]

As we see Israel fail, access to water comes and goes, while bread continues to feed them daily. The reader comes to understand this new role of scarcity in Sinai. Namely, *designed scarcity* is met with divine provision in the Sinai wilderness to act as preparation for the problems of scarcity Israel will face in "the land of milk and honey," where rain is rare and Sabbath is mandatory.

SABBATH AND RAIN: THE LAND OF MILK AND HONEY?

God brings Israel into Canaan. It is a narrow coastal land bridge, a strip right in the middle of the so-called Promised Land—that worldwide territory promised to Abraham, Isaac, and Jacob: "from the river of Egypt to the great river . . . Euphrates" (Gen 15:18). Notably, the boundaries of that Promised Land are rivers. If the river in Egypt refers to the Nile, then the extent of the Promised Land spans westward and eastward to the two seasonally flooding river systems that supported the sprawling empires that hemmed in Canaan.[3]

---

[2]Fretheim, *Exodus*, 187.
[3]Some believe that "the river of Egypt" refers to the western border of Israel: the Wadi El-Arish. Most ancient Jewish sources, from the Targums to Rashi, identify the *nakhal mitsrayim* as the Nile River itself (Josh 15:4). In Gen 15:18, it refers to the *nahar mitsrayim* on the one side and the *nahar gadol perat* (i.e., "the great river Euphrates"). Jeremiah 46:6-8 also juxtaposes the Nile River and the Euphrates.

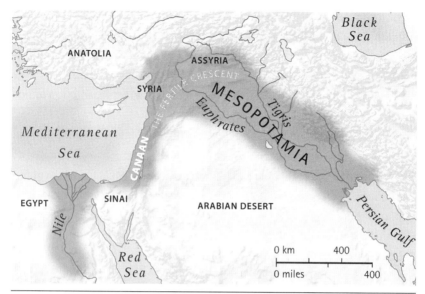

Figure 6.1. The land of Canaan and its neighbors

Oh, the irony. Canaan had few sources of water. Much of Canaan depended entirely on springs, on which cities were built. Otherwise, they relied on catchment water systems where the seasonal rains had to be diverted and trapped in cisterns. These rains came in two months of the year. Farming techniques specific to this largely unarable dirt enabled Israel to grow enough food to survive the year.

Upon settling Canaan, the terrain and climate shaped Israel as Yahweh's people. Because of these, Israel's Canaan-shaped idolatries became God's bone of contention with them. Most of us would be hard-pressed to name Egypt's fertility gods. We would equally be hard-pressed to name anything but the fertility gods of Canaan and its neighbors, Baal and Asherah being the most notable.[4] Knowing the topography and meteorology of Canaan helps us understand why fertility and the gods associated with it feature so prominently in its sacrificial systems.

But isn't Canaan called the "land of milk and honey" in the Torah? That title comes prior to the Iron Age Hebrews. A third-millennium BCE

---

[4]In Egypt, Min was a primary god of fertility, but many renowned gods such as Amun, Osiris, Isis, and Tefnut had fertility as a secondary capability. In Mesopotamia, Inana/Ishtar/Astarte was a more recognizable fertility god.

Egyptian poem refers to Canaan (called Yaa here) similarly to the spies' report in Number 13: "Plentiful was its honey, many were its olives; all manner of fruits were upon its trees. . . . And there were made for me many dainties, and milk prepared in every way" (Tale of Sinuhe, ca. 2100 BCE).[5] We might mistakenly read a breadbasket-like abundance into that description. The phrase "milk and honey," or an equally likely translation, "fat and honey," must be reckoned to the land and water conditions described above.[6] Some scholars have viewed this phrase as a "hyperbolic metaphor." Etan Levine argues that "milk and honey" do not refer to the foods: "Both milk and honey are not products of cultivated farmlands, but of uncultivated grazing areas. The flocks and herds feed on wild growth, on lands unsuitable for agriculture. And it is there, amid the thickets, bushes and wild flowers, that the wild honey is also found."[7] Paired together, "milk and honey" refer to grazing land and thus could be referring to anywhere of decent pasture. Semen is often referred to as "milk" in the ancient Near East, so some think there are hints of sexual fertility within the phrase too.

Others have suggested that *milk* refers to Israel's ability to sustain themselves through pastoral work, and *honey* refers to the naturally occurring food in the land (e.g., Gen 43:11; Judg 14:8; 1 Sam 14:26; etc.). Although archaeologists long thought that beekeeping did not exist in ancient Israel, apiaries recently discovered at Tel Rehov show that bee honey production existed.[8] Taken together, both milk and honey come from areas that have not been settled and worked.

No matter what explanation for "a land of milk and honey" we feel most comfortable with, all explanations fail to imagine what seems to be in mind. Levine calls this the "topographic error" of thinking this land was agriculturally fertile and bountiful.[9]

---

[5]A. H. Gardiner, "Notes on the Story of Sinuhe" (Paris: Librairie Honoré Champion, 1916).

[6]Idan Dershowitz argues convincingly that "fat and honey" makes more sense of the unpointed Hebrew text. Dershowitz, "A Land Flowing with Fat and Honey," *Vetus Testamentum* 60 (2010): 172-76.

[7]Etan Levine, "The Land of Milk and Honey," *Journal for the Study of the Old Testament* 87 (2000): 43-57, here 44.

[8]Amihai Mazar and Nava Panitz-Cohen, "It Is the Land of Honey: Beekeeping at Tel Rehov," *Near Eastern Archaeology* 70, no. 4 (December 2017): 202-19.

[9]Levine believes this language is rhetorically tied to indictments against Israel's misbehaviors in Canaan ("Land of Milk and Honey," 44).

In Ugaritic tablets (written in cuneiform script) dating back to 1400 BCE, Baal is associated with this task of bringing the rainclouds. In this text, Baal is called on by his powers over meteorology:

Listen to me, almighty Ba'al
   Hear me out, Rider of the Clouds.[10]
Almighty Ba'al becomes frightened,
   The Rider of the Clouds is terrified.[11]

Ba'al is even called on to rain oil and honey on the land:

Let the heavens rain olive oil,
   Let the dry stream beds flow with honey.
Then I will know that Ba'al the almighty lives,
   I will know that the most high lord of the earth breathes.[12]

So it is not surprising that across the Hebrew Bible, "milk and honey" describes a land where folks can survive if Yahweh their God delivers the rains and they do not turn to Baal in their anxieties about delayed rains.

All of this leads Philip Stern to the conclusion: "The expression, 'a land flowing with milk and honey' should not be viewed as stereotyped or as a lesson in simple pastoral economics, but as evidence of struggle, a Yahwistic counter-slogan, as it were, in the continuing battle to attach Israel to Yahweh and to ward off the attractions of arch-rival Baal."[13] In the end, two practices of Israel will differentiate the Hebrews' view of scarcity from that of their neighbors: Sabbath and Yahweh as the god of rain. Deuteronomy 8 reminds Israel that the wilderness acted as a scarcity test of their dependence on God:

Remember the long way that the LORD your God has led you these forty years in the wilderness, in order to humble you, *testing you* to know what was in your heart, whether or not you would keep his commandments. *He humbled you by letting you hunger*, then by feeding you with manna, with which neither you nor your ancestors were acquainted, *in order to make you understand* that

---

[10]Victor H. Matthews and Don C. Benjamin, eds., *Old Testament Parallels: Laws and Stories from the Ancient Near East* (Mahwah, NJ: Paulist Press, 1997), 250.

[11]Matthews and Benjamin, *Old Testament Parallels*, 252.

[12]Matthews and Benjamin, *Old Testament Parallels*, 254.

[13]Philip D. Stern, "The Origin and Significance of 'The Land Flowing with Milk and Honey,'" *Vetus Testamentum* 42, no. 4 (1992): 554-57.

one does not live by bread alone but by every word that comes from the mouth of the LORD. (Deut 8:2-3 NRSVUE)

Sabbath also seemed foolish in antiquity. Even up into Hellenistic Judaism, Romans disdained the Jewish Sabbath as laziness and only accommodated it for market-driven reasons.[14] In a subsistent agrarian homestead in Iron Age Israel, stepping away from one's flocks, fields, and tools for an entire day per week would appear, among other things, foolish. Practicing Sabbath would not normally comport with ensuring a family's livelihood year in and out. Too many things could go wrong among too many things that required daily tending. Unlike our modern households, Israelite homesteads did not basically manage themselves day to day.

Because Baal was the god who brought the rains and fostered fertility, temptations to serve food offerings to him and his female consort, Asherah, abounded in the parched sections of Israel. Scrawled across the story of Israel's entry into Canaan, the primary mechanisms for revealing and reifying trust in God are Sabbath practice and rain-trusting subsistence. Israel cannot turn to local fertility gods when the seasonal rains are slow in coming. The biblical narrators of Israel's history code scarcity of water and food into the settled land of Israel as opportunities to see their God provide. That provision is tested weekly (Sabbath), annually (rains), and septennially (seven-year land rests) according to the festival cycles of the Torah.

It is not surprising, then, that Israel's prophets later excoriate Hebrews who fail to practice Sabbath as they celebrate foreigners, even eunuchs, who keep it (see Is 56:2, 4, 6; 58:13; 66:23; Jer 17:21-27; Ezek 20:12-24; 22:8; 23:38; Neh 13:15-22).[15] Sabbath appears to be a weekly practice of trust that funds the greater trust in God needed to live fruitfully in this fragile "land flowing with milk and honey."

The land is a "good land," but *good* here means something more complex than abundant. The land can only be good when Israel trusts Yahweh for the promised rains and lives as a morally just community. Deuteronomy 11 flags the difference between irrigated lands such as Egypt and Mesopotamia with

---

[14]The Roman Stoic Seneca is claimed by Augustine to have sneered at Sabbath as wasting one-seventh of one's life with idleness. Seneca, *On Superstition*, preserved in Augustine, *The City of God Against the Pagans* 6.11.

[15]See also, on hypocritical Sabbath-keeping, Hos 2:11; Amos 8:5.

Canaan's promise of milk and honey. It is a long quote, but worth reading carefully.

> For the land that you are about to enter to *occupy is not like the land of Egypt*, from which you have come, *where you sow your seed and irrigate by foot like a vegetable garden*. But the land that you are crossing over to *occupy is a land of hills and valleys watered by rain from the sky, a land that Yahweh your God looks after*. The eyes of Yahweh your God are always on it, from the beginning of the year to the end of the year.
>
> If you will only heed his every commandment that I am commanding you today—loving Yahweh your God and serving him with all your heart and with all your soul—*then he will give the rain for your land in its season*, the early rain and the later rain, and you will gather in your grain, your wine, and your oil. (Deut 11:10-14 NRSVUE modified)

The land is not *good* in the bland sense of the word. Rather, it can be productive under the ministerial rains brought by Yahweh, not Baal. God's care of the land, Israel's cultivation of justice, and maintenance of proper rituals of Sabbath and sacrifice are the strangely combined prerequisites for this hilly and rocky land to be considered "good."[16]

Hence, the offering of firstfruits from the ground in Deuteronomy 26 combines Israel's moral behaviors and the produce of the ground given by God: "I have obeyed Yahweh my God, doing just as you commanded me. Look down from your holy habitation, from heaven, and bless your people Israel and the ground that you have given us, as you swore to our ancestors, a land flowing with milk and honey" (Deut 26:14-15 NRSVUE modified). We can now see that the phrases "good land" and "flowing with milk and honey" have a thick moral and ritual girth to them rather than referring to a laissez-faire garden that autogenerates sufficient calories for survival.

CONQUEST, KINGS, AND THE COLLAPSE OF AGRICULTURE

Leviticus and Deuteronomy sketch out the possibilities of a new nation emerging from Egypt. The landscape of possibility they portray is dotted

---

[16]Thanks to Michael Rhodes for pointing this out to me.

with Torah-attentive Israelites proliferating in Canaan. Like the garden at Eden, Israel is pictured as fruitful and secure:

> You shall *keep my Sabbaths* and reverence my sanctuary: I am Yahweh.
>
> If you *follow my statutes and keep my commandments* and observe them faithfully, *I will give you your rains in their season, and the land shall yield its produce, and the trees of the field shall yield their fruit.* Your threshing shall overlap the vintage, and the vintage shall overlap the sowing; *you shall eat your bread to the full, and live securely in your land. And I will grant peace in the land,* and you shall lie down, and no one shall make you afraid; I will remove dangerous animals from the land, and no sword shall go through your land. (Lev 26:2-6 NRSVUE modified)

The fragile land, both geopolitically and agriculturally, can be fertile and tranquil or barren and menacing depending on Israel's relation to their neighbors, foreigners, fields, flocks, and God himself.

The same is repeated in Deuteronomy's blessings for following the covenant's instructions (Deut 28:1-12). Because everything listed as a potential blessing here can also be turned *in reverse* as a curse in case of Israel's future oppression of the poor and idolatry (more on this below).

Entering the land of Canaan, the narrator of Joshua is careful to note that the next generation of Israelites trustingly moved across the Jordan River. When they crossed, "on that very day, they ate of the produce of the land [of Canaan]," and the manna ceased the next day (Josh 5:11-12 NRSVUE). The Eden-esque blessings of food depend on Israel's behavior, which makes sense of the service-focused treaty renewal at the end of Joshua.

That renewal at the end of the semiconquest of Canaan cites "this good dirt that Yahweh your God has given you" (Josh 23:13 NRSVUE modified). Joshua also notes, "You eat the fruit of vineyards and olive orchards that you did not plant" (Josh 24:13 NRSVUE modified), as evidence of God's promised fulfillment. The focus is on their ability to tend to God's instruction, not just their crops. The comedy of errors that follows this treaty renewal in Joshua 24 spectacularly illustrates its morbid concerns. Israel will respond to scarcity with violence.

SCARCITY, FAITHLESSNESS, AND WAR

Two touch points can illustrate how Israel's errors and their scarcity cycle go together in the history of Iron Age Israelites. The first form of scarcity stems from the faithlessness of Israel and the second from war, which reveals the faithlessness of Israel. Throughout Israel's divided kingdom (ca. 1000–586 BCE), we trace scarcity and war as they emerge.

The book of the Kings opens with a morally questionable but resplendent King Solomon. Among other fascinations, God endows him with wisdom discernibly portrayed in how he handles the two women with one living infant (and one dead infant). All of Israel hears about Solomon's wisdom and "stood in awe of the king" (1 Kings 3:28 NRSVUE). Paradoxically, Solomon eventually leads Israel to worship other gods, even by sacrificing children. He dies ignominiously, and the nation divides between worse and idolatrous kings of Israel and Judah.

Now, in the world of a kingdom divided, God has sent the mean scarcity of famine to punish the northern nation of Israel for its rampant idolatry in worshiping Baal. King Ahab's idolatry derives ostensibly from his marriage to the foreign royal Jezebel. God tells Elijah that he will invoke famine by withholding the rains, thereby demonstrating who has authority over fertility: God through Elijah, not the god of rain, Baal (1 Kings 17–18).

Fast-forward to the time of Elisha (2 Kings 6), Samaria in the north faces yet another famine and war from invading Syrian forces (2 Kings 6:25). Just as the curses of Deuteronomy 28 horrifically promised, cannibalism ensues:

> It shall besiege you in all your towns until your high and fortified walls, in which you trusted, come down throughout your land; it shall besiege you in all your towns throughout the land that the LORD your God has given you. In the desperate straits to which the enemy siege reduces you, you will eat the fruit of your womb, the flesh of your own sons and daughters whom the LORD your God has given you. (Deut 28:52-53 NRSVUE)

Later in the book of Kings, we see two women fighting over two children: one they have already killed and eaten, and the other still alive. The women who plotted to kill and eat both of their children now seek wisdom from an anonymous king of Israel. The king, now distraught from the horrific absurdity, can only inappropriately focus his rage at the prophet Elisha, who has caused the famine (2 Kings 6:31). Here we see a matrix of anti-Torah unethical behaviors stewing in the cauldron of famine-as-divine-punishment: price gouging (2 Kings 6:25), cannibalism (2 Kings 6:28-29), and failure of leadership. Scarcity begets violence, which begets scarcity, which begets violence.

The second touch point occurs in texts such as Hosea, which hark back to the treaty's descriptions of Edenic provision and security. In Hosea's case, the charge is against Israel, who takes the fruit grown by God's rains and then offers them to Baal. Notice the "naked" and "shamed" references, and that God specifically creates scarcity—wilderness and dehydration—in order to redirect Israel:

> *I will strip her naked*
>> and *expose her as in the day she was born,*
> and *make her like a wilderness,*
>> and *turn her into a parched land,*
>> and *kill her with thirst . . .*
> For she said, "*I will go after my lovers;*
>> *they give me my bread and my water,*
>> *my wool and my flax, my oil and my drink."* . . .
> She did not know
>> that it was *I who gave her*
>> *the grain, the wine, and the oil,*
> and who lavished upon her silver
>> and gold *that they used for Baal.*
> Therefore I will take back
>> my grain in its time,
>> and my wine in its season;
> and I will take away my wool and my flax,
>> *which were to cover her nakedness* ['erwah].
> *Now I will uncover her shame* [nabal]
>> in the sight of her lovers,
>> and no one shall rescue her out of my hand. (Hos 2:3, 5, 8-10 NRSVUE)

But Hosea points out that it did not have to be this way. The future could behold the Edenic mercy of provision and security from *the same dirt* they fruitlessly toil over in times of famine:

I will be like the dew to Israel;
   he shall blossom like the lily;
   he shall strike root like the forests of Lebanon.
His shoots shall spread out;
   his beauty shall be like the olive tree,
   and his fragrance like that of Lebanon.
They shall again live beneath my shadow;
   they shall flourish as a garden;
they shall blossom like the vine;
   their fragrance shall be like the wine of Lebanon. (Hos 14:5-7)

But present realities of Israel's whoredom eventually cause later prophets to delay their visions of bounty and security to an ambiguous future—the day of Yahweh yet to come.

PROVISION AND SECURITY IN THE END

Surveying the biblical literature, we have seen how the tacit acknowledgment of scarcity and insecurity in Eden dissolves the garden's provision and safety. After the dirt's curse of agricultural ruin and exile, which creates vulnerability, we see violence and scarcity coupled (e.g., Cain and Abel, Jezebel and Naboth, the two cannibal mothers, etc.) but immediately teased apart: scarcity does not entail violence, and such violence will be judged separately. Scarcity should turn Israel to God, not to violent solutions in the hands of humans.

God's treaty with Abraham guarantees the patriarchs a level of nomadic provision not afforded to the later generation who left Egypt and arrived at Mount Sinai. God tests Israel at Sinai with scarcity and brings them into a fragile land with the promise of either provision or scarcity based on their treatment of the soil, flocks, families, fields, and foreigners (Deut 28:1-24).

As the prophets turn to the future release from the curses on the earth, they appeal to the same themes of provision and security to depict the days of judgment, both near and final. Four aspects of the prophets' oracles reinforce the relation between scarcity and violence as we have observed it.

First, idolatry is a reaction to scarcity, and scarcity mixed with violence is God's reaction to idolatry. When anxious about the much-needed rains not coming, Israel turns to other specialized gods for fertility. Jeremiah, Ezekiel, and Revelation portray God's judgment against Israel's idolatry in the formulaic promise of sword, famine, and pestilence. They receive the very thing they fear most. They receive the precise opposite of provision and security. Due to Israel's wickedness, they receive violence and oppression against vulnerable folks, which was enabled by idolatry in times of famine (e.g., Jer 24:10; Ezek 14:21).

And when Israel asks, "Why has Yahweh pronounced all this great evil against us?" Jeremiah reminds them of their idolatry, greater than their ancestors (Jer 16:10 modified). It is important to reiterate the presumption that Israel's idolatry aims at local fertility cults, presumably driven by anxieties about the lack of rain or dwindling resources. Idolatry can be, among other things, an inappropriate and habituated reaction to scarcity. The agrarian punishment, like that given to Egypt in Exodus, is reciprocal. Where scarcity once tested Israel to reveal their trust in Yahweh, God pairs famine and violence as judgment after centuries of the absent trust and exploitative violence against the weak that idolatry seems to entail.[17]

Second, biblical authors offer portraits of Edenic provision and security in Canaan. The language of fruitfulness and security is repeated to describe Israel's life in Canaan, but only if Israelites follow Yahweh's guidance. If not, everything that could have been fruitful will be metaphysically reoriented against Israel for frustration (Deut 28:15-68). Deuteronomy's treaty renewal allays anxieties about scarcity in Israel's obedience but warns of punishment for polis-wide injustice in terms of horrifying scarcity and violence.

In a poetically gruesome twist on scarcity and security, both men and women will secretly kill and eat their own children, not sharing the meat with spouses "until [their] high and fortified walls, in which [they] trusted, come down throughout [the] land" (Deut 28:52 NRSVUE).

Third, Isaiah alerts Israel to God's tests of scarcity in order to heighten a future expectation of freedom from it. Where God previously used the "bread of adversity and water of affliction," God will "give rain for the seed with which you sow the dirt, and grain, the produce of the dirt, which will be rich and plenteous" (Is 30:20, 23 NRSVUE modified). Likewise, Amos,

---

[17]As Matthew J. Lynch demonstrates, one of the hallmarks of the grammars of violence in the Hebrew Bible is the victim crying out or the implication that God has heard them cry and has seen what Israel does in secret. Lynch, *Portraying Violence in the Hebrew Bible* (New York: Cambridge University Press, 2020), 151-66.

Zechariah, and other prophets depict the return from exile in terms of safety and bounty: "For there shall be a sowing of peace; the vine shall yield its fruit, the dirt shall give its produce, and the skies shall give their dew, and I will cause the remnant of this people to possess all these things" (Zech 8:12 NRSVUE; see Amos 9:11-15).

Fourth, Edenic provision and security, though never realized in Canaan, become the palette for illustrating the new heavens and earth to come—both in the Hebrew Bible and New Testament. Isaiah's sketch of the times after the final judgment features both plant and animal fertility (Is 65:17-25). It is a place where infants and the elderly do not die. This new heavens and earth conflate Eden and the eschaton through an inconceivable animal society where dust will be the serpent's food, but the "wolf and the lamb shall feed together" (Is 65:25 NRSVUE).

Jesus' parables of his own return often feature feasts in security, though it will be signaled by cosmic-scale disruptions such as earthquakes, wars, famines, and destabilizing rumors of more (Mt 24:4-8; Mk 13:5-8; Lk 21:8-11). Most basically, the "good news" itself is characterized by a parable about crop fertility (i.e., parable of the sower, Mk 4:1-9) or a flourishing mustard bush, and then juxtaposed against a barren fig tree of Israel (i.e., Lk 13:6-9). The Last Supper, which reritualizes at least one festival meal (Passover), is countered with a kingdom coming after Jesus' return that can naturally be depicted as a banquet feast (Lk 14:24).

Second Peter and John's Revelation encourage hope in this same "new heavens and a new earth" (2 Pet 3:13). The final scenes of John's apocalyptic visions present a heaven-earth fusion without pain or fear (Rev 21:1, 4) and a city of fruitful trees divinely fed by rivers of life (Rev 22:1-4).

A consistent pulse emerges across the Scriptures of *what Eden was, what Canaan could have been, and what the new heavens and new earth will be.* God planned good scarcity for Israel. Jesus did the same when he sent his disciples out empty-handed. It is a tactic aimed at revealing trust and morality. This tactic also appears predicated on the hope that anxieties about security and scarcity melt in the glow of a divinely provided and secured realm meant to be spread out to the nations by Israel. But in the light of Israel's failures to become the Israel who lives out the justice of the Torah, Jesus inaugurates this eschatology to be brought later in toto by his return.

CHAPTER SEVEN

BETWEEN EVOLUTIONARY SCARCITY AND SCRIPTURE'S PROVISION

DO EVOLUTIONARY SCIENCE and the biblical literature carry out two identical discussions of scarcity and violence? Though complicated, the short answer is: sometimes. The biblical authors focus and sharpen their beams on the relation between scarcity and violence from beginning to end. Some recent efforts in evolutionary thinking depict cooperation in hominin development, but also more broadly in bacteria, trees, and animals. For both, nature is not naturally or necessarily "red in tooth and claw." Under these explanations in the evolutionary sciences, violence is separable from scarcity in natural selection.

For the biblical authors, provision cannot be manufactured, even though it can be coordinated between God and humanity (e.g., through Joseph's faithfulness, through Israel's behaviors, etc.). Yet, even today, provision and superprovision of food has *not* solved the problem of competition for scarce resources through violence, food waste, or disparate access to food. Hasn't superabundance of resources even exacerbated dictatorships and violence in recent experience? The widespread problem of scarcity in sub-Saharan Africa is what the Peace Corps sought to remedy. Yet when Paul Theroux wrote about his own return to Africa forty years after energetically joining the Peace Corps, he saw how the work of Western volunteers often

enabled dictatorships of various kinds to exploit human and natural resources across Africa. In short, the United States' fight against resource
scarcity seemed to enable those who would exploit the vulnerable.[1] The
biblical discussion of scarcity, violence, and exploitation of the vulnerable
not only anticipates such realities but warns against stratifying the forces
that foment them.

Across Scripture, scarcity fosters and reveals trusts in either Israel's God
or other gods tasked with fertility. In the biblical discourse, scarcity is not a
necessary catalyst for violently competitive relationships, and neither are
cooperative relationships that can generate genetically advantageous features.

To summarize what points I now take to be obvious from the biblical
literature pertaining to the relationship of scarcity to violence:

- Scarcity is a latent concern in the Eden narrative (e.g., "naked" and
  "not ashamed" before exile occurs)—a possible-but-unrealized state
  of humanity until east of Eden.

- Sufficient provision is normative, *nakedness without shame* hints at
  social and physical security, but what will later be described as poverty
  in the Bible is not a necessary state of affairs.[2]

- Scarcity is an ever-present anxiety for those who depend on the
  cursed dirt.

- Violence and scarcity are laced together from Cain to Noah, where
  violence is depicted as nonnecessary and separable from scarcity, but
  within the ken of individual desire and self-control in the first story of
  violence (Gen 4:7).

---

[1]Paul Theroux, *Dark Star Safari: Overland from Cairo to Cape Town* (Boston: Houghton Mifflin
Harcourt, 2003). The Food and Agricultural Organization of the United Nations reports,
"Every year, consumers in rich countries waste almost as much food (222 million tonnes) as
the entire net food production of sub-Saharan Africa (230 million tonnes)." See "Cutting Food
Waste to Save the World," Food and Agricultural Organization of the United Nations, May 11,
2011, www.fao.org/newsroom/detail/Cutting-food-waste-to-feed-the-world/en.
[2]Poverty, in the Hebrew Bible and New Testament, appears in a collection of circumstances not
easily defined. Domeris clarifies the difference between persons facing scarcity and poverty:
"This means that the semantic domain of poverty is tied to several other domains, including
wealth, power, honour and righteousness. To limit the biblical understanding of poverty to
economics is to fail to hear what the Bible is saying about the different dimensions of poverty,
as this book will make clear." William Domeris, *The Social Construction of Poverty Among
Biblical Peasants*, Library of Hebrew Bible/Old Testament Studies 466 (New York: T&T Clark,
2007), 26.

- Ultimately, in Scripture, scarcity is a communal matter and belongs to God's desire to provide and his willingness to judge injustices— necessarily mitigated by living as a specific kind of community, not individual, under God's guidance.

A PERSPECTIVE FROM THE EVOLUTIONARY SCIENCES

In Darwin's view, and much of the evolutionary thinking that developed in his wake, scarcity is normative, and violent competition sometimes emerges as a consequence. Cooperative models of evolution have recently argued for biological altruism, if that is the correct term, which creates a rift in the competition-rich evolutionary story of origins.

For Darwin, where violence occurs, it stems from the biological order and can rarely be understood as volitional. The competition for resources in the struggle for life does not evaporate in the evidence of cooperation but only complicates the story. Violence, in some cases, cannot be helped, but it also should not be stymied. Like seasonal fires in a deciduous forest, competition and violence can be healthy for the long-term development of the species.

The absence of such violence can be bad, relatively speaking. For example, the predation of wolves recently changed the heights of trees, increased the population of birds and beavers, renewed forests, and eventually stabilized the course of rivers in Yellowstone National Park. The sudden cascade of changes to the ecosystem (called "trophic cascade") began with the reintroduction of wolves that preyed on elk and deer.[3] The story of violent competition in Yellowstone shows that some ecosystems benefit greatly from the cascading effects of predation. Cooperation through competition occurs across flora and fauna in unexpected ways.

Our two origins stories could overlap if we imagine an Eden-like cooperative model of human evolution akin to that which the cooperative models suggest. It might go something like this:

> Bonobo-like hominins in a secure and abundant forest would employ nonviolent means to reconcile the myriad conflicts that inevitably arose. Violence,

---

[3]William J. Ripple and Robert L. Beschta, "Trophic Cascades in Yellowstone: The First 15 Years After Wolf Reintroduction," *Biological Conservation* 145, no. 1 (2012): 205-13. Thanks to Taylor Lindsay Dyck for this tip.

when it erupted, would be entirely out of sorts with the cultivated normalcy of reconciliation, so much so, that the eventual problems of scarcity would not inevitably or "naturally" lead to violent competition. Through culturally imbued mutuality, they would navigate scarcity peaceably.

Recorded human history has not always followed this cooperative path, which supplies us with anecdotal reasons to be suspicious of cooperation as an explanatory panacea. I think most evolutionary scientists would agree with this caution. The biblical authors, as well, want us to be suspicious of such accounts, especially where they might depend on no divine accountability for cooperative relationships in the face of famine.

I cannot assess whether this vision of hominin evolution maps onto the history of humanity or naively portrays it as cooperative. I am more concerned here to notice that the biblical authors, like most of their ancient peers, did not conceive of human development this way. For them, a rebellion metaphysically corrupted the earth to create scarcity. This superscarcity gets bound together with calamitous violence requiring a divine filicide by Yahweh against his children. He kills them all, except one family.

Throughout and beyond this discourse on scarcity, generation, and violence in Genesis 1–11, the separation of violence from scarcity depends on trusting Yahweh to provide. The remedy to scarcity is neither abundance of food nor altruism between the starving, but sufficient provision by Yahweh.[4] That divine sufficiency is then strategically configured into a system that requires the sharing of resources (e.g., Deut 15:7-11) and lifting the hungry Hebrews out of their poverty, as they would have a foreigner (Lev 25:35-38).

SWAMIDASS'S GENEALOGICAL ADAM AND EVE

An alternative proposal has been put forward by Joshua Swamidass. His theory of genealogical descent accommodates evolutionary processes that contain both violence and cooperation in scarcity and abundance. In this view, the garden at Eden was the divinely curated and protected exception to the biologically informed developments outside the garden. This also

---

[4]Thanks to Mary Vanden Berg for pointing out that *abundance* is not quite the right word to describe Yahweh's provisions in Eden or later for Israel.

gives Swamidass all the constituent pieces for the biblical trajectory I just previously outlined.[5]

Swamidass—a computational biologist—contends that there is a scientifically realistic account of human history that would have Adam and Eve being among the genealogical ancestors of all living humans in the first century AD. Even if they lived just four thousand years before Jesus, Adam and Eve would still be genealogically related to everyone by the time he walked the earth. Still, we cannot genetically trace our relationship to most of our ancestors. So, we can't see genetic relationships between specific individuals more than a dozen generations back. Prior to that, too much genetic information is lost for us to have any chance of seeing Adam and Eve in our genomes.

Despite this, Swamidass shows how we can plausibly map out genealogical relationships, demonstrating the possibility of generations spanning thousands of years. After running computer simulations of such possibilities, the so-called bottleneck of humans needed to form our current DNA is not in the thousands or dozens, as has been commonly speculated. Rather, the models reveal that it could be one man and one woman, who lived a few thousand years before Jesus, intermixing with an already existing evolved population of hominins outside the garden at Eden. If this were the case, Swamidass maintains that we could have a real couple to whom every human being on earth in the time of Jesus really could be genealogically related.

His point is that the mathematic models and genetics do not preclude the possibility of one kind of literalistic reading of Genesis 2–3. The contention with Swamidass's view hangs on how universal in scope the biblical discourse intends to be in its descriptions of creation. Does Genesis describe every human's generation as necessarily and uniquely stemming from Eden? Is the evolution of hominins outside and prior to Eden a viable history of humanity with which the biblical authors could dovetail their accounts of creation?

If the biblical account could be reconciled to Swamidass's view, I hope that we can now see that it is not merely the biblical history of creation that

---

[5]S. Joshua Swamidass, *The Genealogical Adam and Eve: The Surprising Science of Universal Ancestry* (Downers Grove, IL: IVP Academic, 2019).

needs to be reckoned to his project. The conceptual world of the biblical authors on matters of evolutionary science also needs resolving. Note what I am doing. I am arguing that the most important thing for us to understand is the biblical authors' deep claims about reality. By simply focusing on natural history, Swamidass has actually sidestepped this (and to be fair, this is not his aim). But by focusing on these metaphysical claims, we discover the conceptual commonalities and divergences that the biblical authors might note.

If the biblical authors were merely interested in the history of creation, their brevity and literary style is perplexing for such a task. Rather, they lay out an intellectual program about the nature of creation and its relations by weaving creation into the story of Israel and the new covenant. And it is their conception of the nature of God-creation relations that appears most at odds with some views in the evolutionary sciences, even if the genealogical history can be made to fit (as Swamidass proposes).

Now that we have seen how scarcity features as an undergirding assumption across the biblical literature, we can make sense of antiscarcity rhetoric into the age of resurrection. But does Scripture care about the question of how creatures came to fit their habitats in ways that allowed them to buffet scarcity?

PART THREE

FIT

CHAPTER EIGHT

EVOLUTIONARY FIT AND LOCATION

DO SQUIRRELS LIVE IN TREES because their bodies fit the tree? Or has the tree shaped the squirrel over time to fit its own needs? Of course, it can be both. The relation of fit to location is more complicated than I realized when I first started writing this book.

■ ■ ■

Over the years, I have become aware that I play a character in my freshman classroom called "Dr. Johnson." This professor character says all kinds of things with hidden motives that do not betray to students what I am actually thinking. (Students often assume they know me or know what I am thinking because they know the classroom performance of Dr. Johnson. I thought the same of all my professors too.) With upperclassmen, I play a slightly different version, and less so with graduate students because they have figured out the scheme by then.

I used to think that I brought my lecture content into the classroom and delivered it to the students. But over eighteen years of college teaching, I now see how the classroom shaped this exaggerated version of me called "Dr. Johnson."

Going back nearly two thousand years, we might ask: Which character portrayed to us as the apostle Paul was closest to Paul's actual thinking? Was he basically a Stoic philosopher adapting his Judaism to Roman schools of the day? Many scholars have argued for something like a Stoic Paul. Maybe this is what King Agrippa is referring to when he tries cooling Paul's jets, "Paul, you are out of your mind; your great learning is driving you out of your mind" (Acts 26:24). Others, including me, argue that Paul is basically a Hebrew who thought, walked, and slept Hebraically. He skillfully adapted his ideas to the language and concepts of the Hellenistic Judaism and Roman philosophies of his time.[1] He put things in a way that could have been successfully heard and partially digested.

All these attempts to suss out the most authentic version of Paul often stall because Paul wrote so differently to the various Christian communities. To the Galatians and Romans, he digs deep into Torah and contemporary Jewish commentary. To the Colossians, he waxes on about the cosmic and metaphysical scope of Christ, if not still concerned for Jewish elements among them. To the Corinthians, Paul speaks of mysteries, philosophies, rhetoric, and the many different forms of disunifying behaviors in the community. What is the unified kernel of Paul in all this diversity?

Abraham Malherbe suggests a simple reason the apostle is so difficult to pin down, that "Paul was part of all that he had met."[2] In other words, Paul affected his social environment, but those communities and experiences also sculpted Paul's thinking. When the trees sculpt the squirrels over generations and vice versa, the lines of demarcation are many, malleable, and blurred.

FITTEDNESS TO HABITAT

In this chapter and the next, I engage the concept commonly called "fit to environment" from Darwin's and the biblical perspective. To be honest, this concept of fittedness is a challenging one to convey. So I ask that you hang in there as we proceed. I promise that it will be interesting but not always obvious.

I will briefly summarize Darwin's view of fittedness and eventually situate his discussion in a larger vista of fit, from the wholesale integration of

---

[1] Dru Johnson, *Biblical Philosophy: A Hebraic Approach to the Old and New Testaments* (New York: Cambridge University Press, 2021), 151-80, 203-25. James Thompson writes, "While one may observe the parallels between Paul's correspondences and ancient letters, his letters do not fit into any category." Thompson, *Apostle of Persuasion: Theology and Rhetoric in the Pauline Letters* (Grand Rapids, MI: Baker Academic, 2020), 22; see also C. Kavin Rowe, *One True Life: The Stoics and Early Christians as Rival Traditions* (New Haven, CT: Yale University Press, 2016), 182-205.

[2] Abraham J. Malherbe, *Paul and the Popular Philosophers* (Minneapolis, MN: Fortress, 1989), 67.

bacteria into our bodies to languages shaped by terrain to adaptive features in a violent competition.

The garden story also describes the conditions of proper fit to environment and to others. Alongside matters of scarcity, Genesis 2–3 begins the story of humanity by carefully portraying the man's intimate and biological connections with the garden's soil, the other animals, and the woman—the only creature with which he could fit as "one flesh."

Though the biblical authors have no explicit view of a genetic fit to environment—neither did Darwin—the concept we might call biological fittedness pervades their thinking. It operates almost as a background assumption. Beyond Genesis 2–3, other stories of fittedness to habitat support the idea that humanity, for the biblical authors, is shaped by habitat. The biblical authors show us how humans also shape their habitats with tools. And, of course, humans are also "shaped by our tools."[3] Like Paul, we are all a part of all that we have met.

CREATURE AND PLACE

If variations useful to any organic being ever do occur,
assuredly individuals thus characterised will have the best
chance of being preserved in the struggle for life.
CHARLES DARWIN, *THE ORIGIN OF SPECIES*

Darwin's characterization of "useful" genetic variations functions as shorthand for the evolutionary principle of fittedness. Variations in individuals are useful specifically because those variations can best use the resources in each habitat.

"Fit" is an apt metaphor taken from lived experience. "If the gloves don't fit . . . ," well, you know. "Fit to environment," therefore, encompasses all of a creature's interactions with a particular location under the concept of habitat or niche. But the picture that has emerged in the last generation of biology complicates any simple notion of fit.

Below, I consider how several interrelated features of "fit to environment" factor into selection. Selection can sometimes be simplistically pictured as

---

[3]Sherry Turkle, *Alone Together: Why We Expect More from Technology and Less from Each Other* (New York: Basic Books, 2012), x.

emerging from genetic variations in a cauldron of survival pressures. Yet, our fittedness to a habitat ranges from our microbial assimilation of our habitats into our bodies to adaptive genetic variations that aid in fittedness. It also includes enculturated biases expressed in mate selection by females of a species.[4]

The idea of fit being directly and easily related to environment no longer holds. Instead, social cultures, microbial cultures, and everything in between direct the nature of selection in any given habitat. This is the same tack taken by the biblical narrative regarding the intersection of culture, biome, and sexual selection, which will be reviewed in the coming chapters.

Adaptive peoples to adaptive places. When we think of genetic variations that might favor a group's adaptation to a place, we might remember famous examples such as the tree patterns on moth's wing or a bird's beak on the Galápagos Islands. These adaptations also became a hiccup for Darwin. He noticed that, at some stage, variation might be rooted in something like female preference rather than pure chance.

To illustrate the complexity of adaptation, symbiosis, and fit to habitat, consider the human settlers of the Negev desert in the Neolithic period. They provide a more recent example of hominin adaptation to the extremes of desert life and the mutual adaptations required by settling there.

In the 1950s, Michael Evenari and his wife moved to the interior of Israel's Negev desert to begin a decades-long experiment farming the desert as the ancients did. Evidence suggests human habitation in the Negev desert starting from 11,000 BCE onward. The desert may have been once more verdant until the Bronze Age (ca. 2100 BCE, when it turned arid and has remained so since).

From historical times forward, Evenari found that the ancient settlers had carefully cultivated the desert, avoiding overirrigation, salination of soil, and erosion, and all the while trapping intermittent sources of water with catchment systems. Evenari writes, "The ancient [Negev] farmer fitted his artificially created agricultural eco-systems into nature and used landscape and topography to his best advantage without damaging his

---

[4]The concept of species is hotly contested among biologists today. I use the term in only the most colloquial sense.

environment."[5] Surely this all required a generational long game, never appreciable in one person's lifetime.

Though we idealize them as resource prudent, later Bedouins failed on these fronts, exploiting resources and cutting off future generations from avenues of food and water security. "They mismanaged the desert," says Evenari. Despite "the great attraction that the Bedouin has always had for the romantically inclined," they present a negative case for fit to environs.[6]

Conversely, Evenari notes that not only did ancient farmers adapt themselves to the desert, but also the desert both adapted *to them* and fostered certain adaptive features *in them*. Because of the extremities and precariousness of desert farming, "man had to develop special qualities, spiritual, intellectual, and social, enabling him to overcome the continuous danger to his existence."[7] Humans learned from the Negev and improved on what they learned.

This tightly woven nexus of adaptive relationships between humans and place occurs within a web of chemical and biological relationships. Evenari's understanding of the biosphere captures much of what we consider biological and cultural matters of humanity's fittedness: "All plants, animals, and men living on the planet earth, as well as the soil on which they live and from which they draw water and nutrients and the atmosphere with which they exchange gases, *constitute one equilibrated unit*, the biosphere."[8] Humans are unique, for Evenari, in that we can disturb that equilibrium intentionally or not, for good and for ill. Of course, most of ancient Israelite culture existed just north of the horticulturally bereft Negev desert. In ways akin to the ancient Negev farmer, Israel's houses, instruments, clothes, calendars, and everything they thought and did were intimately shaped by their soil, seed, and water relationships.

Reflecting on those Neolithic Negev farmers, it seems obvious to us that they were shaped *by* and *toward* their environment. Because of the extreme habitats conjured in our imaginations by the "out of Africa" thesis of human origins, it might be difficult for us to conceive of the history of hominins not

---

[5]Michael Evenari, Leslie Shanan, and Naphtali Tadmor, *The Negev: The Challenge of the Desert* (Cambridge, MA: Harvard University Press, 1971), 325.

[6]Evenari, Shanan, and Tadmor, *Negev*, 326.

[7]Evenari, Shanan, and Tadmor, *Negev*, 325.

[8]Evenari, Shanan, and Tadmor, *Negev*, 324, emphasis added.

shaped by extremes of predation and climate. Our evolutionary imaginations often envision a swelter of pressures—arid climate, water and food
scarcity, dangerous flora and fauna—within which some genetic variations
happen to create advantages.

Biologists think something like the Negev farmers might be a better
image to conjure for evolution, where they fit to the habitat full of existential
anxieties and pressures but also created cultures that could shape the habitat.
By doing so, they inevitably shaped their own genetic path forward. Chance
might initiate some changes; but beaks, wings, and tail plumage do not
emerge and form from pure genetic chance. They change in relation to the
niche to which they best fit.

As we will see, the biblical authors also noted the intricate and intimate
connection to and *shaping by* humanity's environs. By such observations,
they prescribed certain ethical obligations to one's habitat.

Human integration with the environment. Though "fit to environment"
generally refers to organisms conforming to the sources of nutrients, to
climate, and to predation of a place, adaptive features can penetrate the
organism as well. That bidirectional relationship between creation and
place appears even more energetic than Darwin could have imagined. Not
only do organisms fit to their environment, but they also fit their environs
to themselves (also called "niche construction") and *into* themselves
(e.g., biomes).

Along with the cultural tailoring of an environment to the organism
(more on this below), complex organisms such as mammals do not just
move in and around their habitats. They assimilate parts of it to function
symbiotically (e.g., helpful bacteria in mammal guts) and asymbiotically
(e.g., the cold virus).

By the numbers, we might question our traditional views of anatomy.
When added up, our human body cells are outnumbered by our microbiome within us by "an order of magnitude" or two or three.[9] Our human
body cells are dwarfed by our bacteria, numerically speaking. And these
are not just any bacterial cells. Our bacterial biome reveals a curated collection of the bacteria our mother passed along to us via her birth canal,

---

[9]Barbara A. Methé et al., "A Framework for Human Microbiome Research," *Nature*, June 13,
2012, 215-21.

the bacteria in the dirt of our elementary school yard, the bacteria we met in the college dorm, the military barracks, and everywhere else we have ever been.

Our DNA reveals edits from the environment as well. We edit snippets of virus-like sequences into our DNA, presumably to give our cells a viral profile to look out for. That to say, aspects of our viral history can be read in the edits of our DNA.

We are all biological curio cabinets. When it comes to our human microbiome, like the apostle Paul, we are all part of all that we have ever met.[10] We have appropriated the microbial world around us throughout our lives. Researchers are currently working on ways to use our distinct biomes to distinguish us like fingerprints or DNA do.[11] Hence, the concept of a human as a microbiome (i.e., a balanced collection of organisms) has become a favored metaphor in research on nutrition, infectious diseases, cancer, and more.[12]

Not only are mammals chock-full of bacteria, but we seem to relate to our environs less intentionally than we might have suspected. Research continues to reveal that the relation between human body and external world is dynamic and complex. The microbiome is merely one matrix of relationships between a mammal's body and its environs.

Studies have also found that not only mammals but even trees "talk" to each other through the soil's fungal networks, communicating their need for nutrients, which can then be exchanged between them.[13] Trees even "speak" to each other and "share" nutrients cross-species. Though it is debated

---

[10]Once thought to be a ten-to-one ratio of bacteria to human body cells, the current constrained consensus is more comfortable saying that human cells are outnumbered, but by how much we cannot yet say. *Microbiome* refers to the communities of "microbes associated with the human body include eukaryotes, archaea, bacteria, and viruses. . . . Our knowledge of these communities and their gene content, referred to collectively as the human microbiome." Methé et al., "Framework for Human Microbiome Research."

[11]See the National Institute of Health's Human Microbiome Project, www.hmpdacc.org/hmp/.

[12]"These studies have revealed causal mechanisms for both microbes within [cancerous] tumours and microbes in other host niches separated from [cancerous] tumours, mediated through direct and immunological mechanisms." Brian Goodman and Humphrey Gardner, "The Microbiome and Cancer," *Journal of Pathology* 244, no. 5 (April 2018): 667-76.

[13]"Trees communicate their needs and send each other nutrients via a network of latticed fungi buried in the soil." Diane Toomey, "Exploring How and Why Trees 'Talk' to Each Other," Yale School of Forestry and Environmental Studies, September 1, 2016, https://e360.yale.edu/features /exploring_how_and_why_trees_talk_to_each_other/.

whether the tree communication and sharing is cooperative or competitive, something is shared between them.[14] The thick lines we might have visualized between a human and her environment, or a tree and the forest, now look thinner and more perforated.

Evolutionary theorists have advocated that the most basic kind of fittedness to environment is not a static genetic reality, such as a strategic camouflage against predation. A moth's camouflage aids in natural selection, but not always in the larger course of evolution. Rather, evolutionary science now thinks about evolution as an interplay between organisms (genetically), their communities (culturally), and their collective response to changing environs (adaptability).

Darwin showed his sensitivities to these complex relationships and appeals to them as both proliferatingly complex and selection-driving realities.[15] Starting with the climate, Darwin thinks of extreme temperatures as the "most effective of all checks" to extinguish ill-fitted lines in his tree of life. To those that survived, he notes the elaborate features inside and out that tip us off to the environment's shaping of a species: "Almost every part of every organic being is so beautifully related to its complex conditions of life that it seems as improbable that any part should have been suddenly produced perfect, as that a complex machine should have been invented by man in a perfect state."[16]

The struggle for sex: Let the ladies decide? The surprising role of enculturated preferences in selection inserts a wrinkle into any smooth account of fit to environment. Current evolutionary science deals with this complex reality of fittedness—exchange, hosting, and symbiosis—at the layers of culture and genetics. These two parallel tracks form and inform selection respectively. Genetics was once thought to be the "hard coding" of our biology, while culture was thought to influence the accidental properties of our lives: food palettes, fur colors, or choice of mate. However, the impact

---

[14]"Scientists have yet to show that these webs are widespread or ecologically significant in forests." Gabriel Popkin, "Are Trees Talking Underground? For Scientists, It's in Dispute," *New York Times*, November 7, 2022, www.nytimes.com/2022/11/07/science/trees-fungi-talking.html/.

[15]"I am convinced that the most experienced naturalist would be surprised at the number of the cases of variability, even in important parts of structure, which he could collect on good authority, as I have collected, during a course of years." Charles Darwin, *The Origin of Species, by Means of Natural Selection*, 6th ed. (London: John Murray, 1872), 34.

[16]Darwin, *Origin of Species*, 54, 33.

of culture on genetics may be more basic and integral to the evolutionary patterns of development.

Darwin was aware of both the genetically adapted features of an organism and cultural influences on selection. He writes, "I am convinced that natural selection has been the most important, but not the exclusive, means of modification."[17] He was sensitive to the subtle complexities introduced by cultural preferences over genetic speciation. For him, the matter becomes obvious in the tail plumage of male birds.

Darwin observed that the male birds of some species tend to have plumage and other features reserved exclusively for mating rituals. These special feathers and features aim at precisely one task: enticing the less-eager female to mate, which, when successful, passes along that male's genes, including his female-selected plumage. The male that appeals to the female most by means of dress, voice, housing, or dance wins. In ways that many of us can identify with today, this is not a struggle for life but a struggle for reproduction (more on this in chapter ten).

The implication of Darwin's rather ordinary observation instigated a debate among his contemporaries. What incensed some of them was the idea that the female's personal preference might determine the shape and color of the species' evolutionary path. The female's many-faceted preferences cannot all be accounted for, and some of them are a product of culturation—certain preferences that arise *not out of necessity* but *desire*, for lack of a better term.

To see the problem more clearly, consider that natural selection is meant to describe a competition that favors one's fittedness to the niche environment. Yet, the male's excessive plumage and female's slighter shape might result from mating practices. The genetic characteristics that develop because of those differences in sexes must be traced back to something. What else could it be other than the female's preference for the accidental plumage or song patter introduced by something like random gene mutations in the males? Moreover, preferences for accidental features such as tail plumage have been shown to pass generationally, female to female. And many of these features that factor in mating have the opposite effect from adaptive strategies in natural selection.

---

[17]Darwin, *Origin of Species*, 4.

They can create males more likely to be killed in the struggle for life. The very feathers that females prefer also might slow a male down and flag him up to predators. This can mean that it is all sex and death for some males.

For humans, consider right-handedness. Kevin Laland argues that handedness is a case of culture overriding genetic selection: "Purely genetic accounts of handedness fail to explain the well-established cultural influences on handedness. Left-handers are found at lower frequencies in societies that associate it with clumsiness, evil, dirtiness or mental illness."[18] Cultural preferences for right-handedness seems to override any genetic distribution of left-handedness.

The same goes for mate selection, though Laland sees evidence from social psychology that genetic selection remains primary: "Even if human mating preferences are learned, socially transmitted, and culture-specific . . . culturally generated sexual selection may be even more potent than its conventional gene-based counterpart." In brief, Laland thinks that "cultural transmission can overwhelm natural selection."[19] (This might also help to explain parachute pants, vests, glitter makeup, and other odd hominin preferences.)

Most view our ability to culturally mutate our genes as a more recent phenomenon in hominin history, but it appears to be part of natural selection (so long as we agree about animals having "preferences" in "nature"). We might think that culturally formed selection marks the distinguishing feature of humanity against all other animals: "Whenever a need arises, humans can directly pursue the appropriate cultural 'mutations,' that is, design changes to meet the challenge."[20] Culturally formed preference can be advantageous. Francis Ayala outlines the benefits of cultural selection as three:

---

[18]Kevin N. Laland, "Exploring Gene-Culture Interactions: Insights from Handedness, Sexual Selection and Niche-Construction Case Studies," *Philosophical Transactions of the Royal Society of London, Series B, Biological Sciences* 363, no. 1509 (2008): 3577-89.

[19]Laland, "Exploring Gene-Culture Interactions," 3585. He goes on to say, "Niche construction is typically functional and adaptive because it is informed, but not determined, by genes, and sometimes also by learning and culture. Humans largely construct their world to suit themselves, leaving human behaviour largely adaptive in spite of the transformations they have brought about in the environment." Laland, 3586.

[20]"For the last few millennia . . . it was the appearance of culture as a superorganic form of adaptation that made humankind the most successful animal species." Francisco J. Ayala, "Human Evolution: The Three Grand Challenges of Human Biology," in *The Cambridge Companion to the Philosophy of Biology*, ed. David L. Hull and Michael Ruse (Cambridge: Cambridge University Press, 2007), 233-54, here 251.

Cultural adaptation is more effective than biological adaptation because (1) its innovations are directed, rather than random mutations; (2) it can be transmitted "horizontally," rather than only "vertically," to descendants; and (3) because cultural heredity is Lamarckian [traits affected by use or environment], rather than Mendelian [traits not affected by use or environment], acquired characteristics can be inherited.[21]

For example, the peahen's preference in a mate appears to have roots not only in her genetics but also in local peahen and peacock culture. The peahens are not captivated by that which arrests most of us when we visit a zoo: the tall arc of peacock tailfeathers. Apparently, peahens do not rank males much by the ornate head plumage or the girth of the males' tail fans. Rather, the peacock who wins is the one whose lower plumage held the peahen's attention the longest.[22]

The peacocks (males) show signs of social awareness on this front, spending only 5 percent of their gazing time looking at potential peahen mates. Most of their male gaze was directed at other males, specifically the fullest part of their plumage.[23] They seemed to be aware that the ladies were looking, even if the guys were looking at the wrong part of male plumage. Like human dating practices and media-shaped notions of beauty or success, the cultural dynamics that precede bird mating shape the genetics that ensue it. Peachicks are products of culturally formed genetics.

Even considering the question of what people consider sexy decade over decades, or country by country, immediately reveals culturally formed selection practices. Humans consider some factors of symmetry and body ratios to be universally attractive, but cultural preferences slip in too. Such a question also helps us to understand why these might be potent but not determinative selection biases.

---

[21]Ayala, "Human Evolution," 234.

[22]Jessica L. Yorzinski, et al., "Through Their Eyes: Selective Attention in Peahens During Courtship," *Journal of Experimental Biology* 216 (2013): 3035-46. My favorite part of this research is the mechanics of it all: "Yorzinski added that one of the most difficult parts of the research was trying to strap the little helmet with the eye tracking monitors onto the birds." Jon Herskovitz, "When It Comes to Peacock Mating, Plumage Size Matters," Reuters, March 15, 2017, www.reuters.com/article/us-usa-peacocks/when-it-comes-to-peacock-mating-plumage-size-matters-study-idUSKBN16M373/.

[23]Jessica L. Yorzinski et al., "Selective Attention in Peacocks During Assessment of Rival Males," *Journal of Experimental Biology* 220 (2017): 1146-53.

As with birds, humanity's social evolution is part of the story as well. Human "cooperation as a type of fitness" is now widely regarded to be necessary in evolutionary processes. Though Darwin held to a more innate aggressivism and competition-funded view of survival, cooperation also fit into the scheme. As Ashley Montagu chides, "Darwin was going to have the cake of natural selection, which he had baked, and eat it too. The flavoring was mostly 'struggle for existence,' but here and there was a sprinkling of the thinnest kind of co-operation."[24]

Fit to environment is not a brute factor in evolutionary processes. It can even include what we might consider ethical behaviors. Fittedness will even depend in part on a cultural preference for cooperation in times of plenty and scarcity.

FITTEDNESS IN DISCUSSIONS ON RACE AND PLACE

Debates about genetic fit raged well before Darwin, and they often turned racial if not racist. Thus, theological discussions about genetic fit to a location have had an ignoble history. By the late eighteenth century, a debate roiled over a central evolutionary question du jour: Why do humans across the globe look and act so differently from one another?

The answers split between the environmentalist and the anti-environmentalists. The environmentalists believed that locale shaped genetic features such as skin pigment, bone structure, language, thinking, and so on. They were environmentalist in the sense that they believed the environment of origin shaped the genetic traits of creatures in that location. Places made the human to fit the environment. Or, "places made races."

The anti-environmentalist claimed, "Climate did not make human varieties; rather, human varieties were made [by God] for different climates."[25] God made various races of humans for different places in the world, so the anti-environmentalist thinking went. Some races were created biologically inferior by God and placed thus.

---

[24]Ashley Montagu, *Darwin: Competition and Cooperation* (Westport, CT: Greenwood, 1952), 96.
[25]David N. Livingstone, *Adam's Ancestors: Race, Religion, and the Politics of Human Origins* (Baltimore: Johns Hopkins University Press, 2008), 58.

At stake in this controversy was whether God made all human races originally and therefore God designed different races for different places. Or conversely, do all humans emerge from one primitive form of humanity and therefore share a genealogical connection with each other that becomes genetically diversified over time and in different soils? I will later show how the Babel narrative wants to address this very issue.

Against the view that God created biologically diverse hominins for different climates, the environmentalists invoked the biblical account in Genesis. For them, Genesis portrayed the basic and primitive biological connections among all humanity as descending from the garden's couple. The moral implications of each position follow, though not consistently in the history of this debate. As Europeans moved into the New World, their encounters with various indigenous peoples forced a different question with disastrous answers: Are the indigenous peoples in the Americas human beings?

Sharing basic human biology (environmentalism) meant that the indigenous Americans should be treated as humans—considered worthy of the gospel and dignities afforded. Polygenesis (i.e., "different races for different places") meant that some creatures might be inferior to others and could be set to servitude like beasts of burden.

Environmentalist thinking was later employed to attack the peculiar institution of the European-North American slave trade. Predictably, a theological conviction in the unity of all humanity did not guarantee a parallel belief in the equality of all humans.[26] Indeed, some environmentalists fought for human unity from Genesis according to the doctrine of *imago Dei*, yet would also agree that some climates and soils shaped less progressed human populations.

These primitive folks, so they argued, would have to be dealt with according to their particular inferiorities. Therefore, environmentalism and shared humanity based on the image of God could also produce justifications for enslaving indigenous peoples as inferior. Though European colonizers shared the *imago Dei* by creation with indigenous peoples in the New World, the environment that shaped the indigenous humans was inferior. Such was the state of the debate prior to Darwin.

---

[26]Livingstone, *Adam's Ancestors*, 64-68.

CONCLUSIONS

Darwin saw fittedness as an ever-improving situation for the species: "The ultimate result is that each creature tends to become more and more improved in relation to its conditions."[27] But we can tend to think of fit as a one-way street, that we fit our environment. Studies have shown that the relationships between creatures and their niches have repeatedly surpassed expectations and can even be vigorous, complex, and bidirectional changes of environment and creatures in tandem.

In the end, *both* the genetic features that allowed fit to an environment and culturally shaped preferences, including the preference to cooperate, inevitably influence any origins story. Starting with our biome, we are not independent biological agents. We organically appropriate the biological realms to which we have belonged—wombs, ponds, rooms, forests, and all. This raises the question of who is fitting to whom, or what, when it comes to our ability to survive in any given environment.

Less passively, animal cultures and preferences sculpt our selection biases, which shapes the course of our symbiotic fit to our environment. It is not just climate and predatory accommodations that lead to the success of a genetic feature. What a girl wants bears selection significance. Survival of the fittest this is not. (Maybe survival of the "fittest" in modern British slang. I will try to avoid using the phrase "survival of the sexiest," but there it is.)

Even at the level of cultures and family units, the Neolithic Negev farmers illustrate something more sophisticated at work in fittedness. At some point, farmers enculturated a multigenerational plan to shape the environment as their task of conservation simultaneously shaped these desert farmers. Cultures of conservation were, of course, not the path of all Negev dwellers, as Bedouins appear to have resisted being shaped by the landscape with a "use up, then move on" strategy.

All this to say that "fit to environment" is a conceptual package that includes creatures' interactions with their locale and ranges from cells to civilization, and everything in between. The Hebraic creation narrative takes a remarkably similar tack on animals generally, but also hominins. Now, I turn to that biblical account to assess whether fit and what we now call genetics factor into the biblical discourse on the creation and spread of humanity.

---

[27]Darwin, *Origin of Species*, 97.

CHAPTER NINE

FIT AND DISLOCATION FROM GARDEN TO NEW EARTH

HOW DO HUMANS FIT to our environment according to the biblical literature? Though Genesis depicts a deep and "dirty" affinity between humans, earth, and animals, it is human behaviors that ultimately determine the fit to the place.

From the prior chapter, we see the need to reckon how we have fit our location to us and how we have shaped our culture and one another over against our environment. We saw that the relationship between hominins and environment gets quickly complicated in light of the emerging sociological factors. Yet, Genesis 1–2 carefully paints a picture of fittedness orchestrated by God that is both physiologically and culturally appropriate, entangling the sociology of human origins with its biology in some surprising ways. East of Eden—unsettled and toilsome—the man and woman, Cain, and later Israel suffered from exiles that specifically disallowed fit to environment—physiologically, culturally, linguistically, and otherwise. It is going to get a little wild here, so hang with me.

HABITATS AND INHABITANTS

It is worth stating what is obvious: Genesis 1 comes out of the corner swinging on the topic of fit. Many scholars see the six-day structure of creation as forming and then filling. Days one through three show God creating habitable spaces: heavens, earth, waters. Days four, five, and six are devoted to emplacing the appropriate inhabitants of each space: celestial bodies and birds to the heavens, beasts and creeping things to the land, and fish to the waters.

Only one comment is necessary here, and it is painfully obvious: creation is created by fitting inhabitants to their proper habitats. As we have said with so many other things, this is not necessarily true of other creation accounts in the ancient Near East. The focus is specific and explicit in the repetitions and day cycles of Genesis 1. The bumper-sticker version: Hebrew creation is about creatures fitting to their created domains—water, land, and sky. But what happens when we turn to the specific locale in Genesis 2? The fittedness rhetoric gets amped up.

DIRT-TO-DIRTLING RELATIONS

Given the environmentalist and anti-environmentalist positions drawn from Genesis (see chap. 8), it should be clear that we cannot make bald theological appeals to *the* unambiguous meaning of a concept in Genesis. Examining the literary development of humanity's fittedness to location allows the biblical texts to speak, even if the implications of what they say are not immediately obvious. Below, I attempt to hear out how the biblical authors develop notions of fit to environment.

Let's now spend a few paragraphs on the basic case for the connectedness of the *dirtling* to the woman and to their habitat. Beginning with the wordplay between the *dirtling* and the *dirt* (*haʾadam* and *haʾadamah*, respectively), the narrator stresses the man's relation to his environs by his constitution and his curses alike. He is not called "the man" (*haʾish*) or even "Adam" in Genesis 1–2. Rather, he is almost always referred to by his title, *the dirtling* (*haʾadam*).

If his title did not emphasize it enough, the narrator twice reminds us that *the dirtling* was taken and formed from the *dirt* (Gen 3:19, 23). The animals were also formed of the dirt, and they had the "breath of life" and were called "living creatures" (*nephesh hayot*) just like the dirtling (see Gen 1:30; 2:19; 6:17). Even though they share dirt and breath, only the man is called *the dirtling*. That title intimately connects him to the animals in a way unlike how he is connected to the plants. But he is also separated from the animals, treated differently by the narrator and by God. He is made to realize this separation when he names all the dirt-formed creations with the breath of life but "could not find for himself an ally fit for him" (Gen 2:20 my translation).[1]

God gives animals and humans alike all green plants to eat (Gen 1:30). God binds himself in covenant to animals and humans alike after the flood, naming the features that unite them: "living creatures" with the "breath of life," who are equally held responsible for the shedding of blood (Gen 6:17; 8:21; 9:5). And though Genesis repeatedly binds the fittedness and fates of humans and animals together, the dirtling is different.

Scholars working on the conceptual world of Genesis sometimes miss this connective tissue between *dirt* and *dirtling* by translating *ha'adam* as "human" or "man."[2] This dirt-man-animal-breath link goes beyond a mere one-off topic of Genesis 2. Genesis maintains the dirtling-to-dirt wordplay throughout its primeval history, unlike some parallel concepts of "human" in the ancient Near East.[3] As previously discussed in chapter five, even the resolution of the flood focuses on cursed dirt and not on humanity's evil hearts (see Gen 5:29; 8:21).

Returning to Genesis 2, an enigmatic quiet exists in the story's declining to explain why this dirtling (*ha'adam*), later called "the man" (*ha'ish*), is fundamentally different from the other dirt-formed creatures. That tension might be relieved by assuming what seems obvious: this creation account in Genesis 2 intends to depict the humans as made "in our image . . . male and female" from Genesis 1 (Gen 1:26-27).

The narrator linguistically connects the woman's identity to the dirtling, who is connected to the other living animals in his *dirted-ness*. How? The repetition of the word "take" (*laqakh*) signals the special relation obsessively highlighted by the narrator:

---

[1] The term traditionally translated "helper" (*'ezer*) is often used in military contexts of allegiance and allies across the Hebrew Bible.

[2] E.g., John H. Walton, *The Lost World of Adam and Eve: Genesis 2–3 and the Human Origins Debate* (Downers Grove, IL: IVP Academic, 2015), 72.

[3] Genesis contrasts with some parallel concepts in the ancient Near East. For instance, in Egyptian, they are not called *siltlings* because they were taken from the silt or *claylings* because they were taken from Mesopotamia's clay. In ancient Egypt, distinctions were made between various kinds of foreigners and "humans," implying that only Egyptians were truly human. See Henri Frankfort et al., *The Intellectual Adventure of Ancient Man: An Essay on Speculative Thought in the Ancient Near East* (Chicago: University of Chicago Press, 1977), 33.

God "takes" (*laqakh*) and "settles" (*nuakh*) *the dirtling* into the garden (Gen 2:15).

God "takes" the side part from the dirtling (Gen 2:21).

God constructs (*banah*) the woman from the side "God had *taken* [*laqakh*] from the dirtling" (Gen 2:22 my translation).

The dirtling calls her "woman" ('*ishah*), "for she was *taken* [*laqakh*] out of man ['*ish*]" (Gen 2:23 my translation).

The dirtling's curse is explicitly predicated on being "taken" (*laqakh*) from the dirt (Gen 3:19).

The dirtling is exiled to "work the dirt from which he was *taken* [*laqakh*]" (Gen 3:23 my translation).

Of the thirty-two instances in the entire Hebrew Bible where God is the direct or implied subject of "take" (*laqakh*), six of them occur in the concentrated space of Genesis 2–3. All six instances include some type of wordplay (*ha'adam/ha'adamah* or '*ish*/'*ishah*) and situate the dirtling in reference to the dirt and dust, or the woman in reference to the man. Presumably, these factors lie behind Joseph Blenkinsopp's assertion that "unlike the animals, the woman is connatural with the man in the closest way possible. She is 'bone of his bone, flesh of his flesh,' which is to say that she and the man share kinship of the most intimate kind."[4]

Beyond fitting to each other sexually and otherwise, an ordering of relationships appears to undergird these interconnections. In other words, the beasts, soil, and humans were not just made of the same stuff but animated and related in a structured relationship with one another. Genesis highlights the soil's fertility in the man's curse. The soil was oriented to bear fruit suitable by the service of humans. What would *suitable* entail here? I can only speculate from its antithesis stated in the curse and depicted repeatedly in the history of Israel: suitable would mean not under duress or "toil" ('*itsabon*, Gen 3:17). The same goes for the woman's fertility and fruitfulness, which could have been without "toil" ('*itsabon*, Gen 3:16).

---

[4]Joseph Blenkinsopp, *Creation, Un-creation, Re-creation: A Discursive Commentary on Genesis 1–11* (New York: T&T Clark, 2011), 71.

The exact metaphysical relation of soil to man remains unclear. But the stories repeatedly reinforce that some metaphysical relationship exists. The biblical authors seemed to think some kind of harmonious relation existed, and it existed in a relation outside mere physical ordering. What can we say? The physical world in Eden was metaphysically oriented toward creaturely favor. The relation had the goal of sustaining an orchardist with reproductive fruitfulness.

That harmonious orientation was cursed in Genesis 3 and became the hinge of God's response to Israel more generally. What could otherwise be fruitful would be redirected as barren and frustrating under curses (see Lev 26; Deut 28). Key for the concept of fittedness: the same physical substances could be turned toward or against humanity, flora, and fauna. But why?

We need to admit to some kind of metaphysical orientation in creation to affirm that which the text of Genesis presents and assumes: a good structure turned against itself. The *dirt* now turns against the *dirtling*. It is not merely humans turning against the earth. The same logic and language are repeated in Deuteronomy's treaty renewal, where Deuteronomy 28:1-14 is reversed in Deuteronomy 28:15-68, merely replacing "blessed be" with "cursed be."

Deuteronomy's blessings are stated briefly, while the curses proliferate into a genre we might best describe as "creative fictive horror." This imaginative horror is used to display the kinds of cannibalistic fervor with which Yahweh will starve them out in their high-walled cities, in which they wrongly put their trust and to which they no longer properly fit.

In case we thought this was describing naturalized universal wisdom about seeking knowledge in the garden that leads to prosperity, the curses quickly convince us otherwise. No natural process singles out a people group for either fruitfulness or destruction. And it is the specificity of the curses that rules out anything but a metaphysical reorientation of the cosmos.

So too with the dirtling and the dirt, the woman and childbirth, the serpent and humanity in Genesis 3. What could have been a fruitful relationship is now metaphysically reoriented for dearth, toil, and pain.

Over the course of the Torah and Gospels, we come to understand that the metaphysics of these curses is what makes all of creation simultaneously cursable yet redeemable. Cursing disorients the course of creation, and blessing reorients it. But these matters of reorientation only matter if humanity is morally and metaphysically fit to their habitat.

Providing contrast, the curses show misfit and dislocation. Agriculture can only offer sustenance if families settle and make roots. But God has already promised the frustration of the seed and soil alongside the tumult of the woman's seed with the serpent's (Gen 3:15).

In the clarifying spotlight of the couple's exile from Eden, we see that the couple no longer fits the verdant dirt of the garden (Gen 3:17-18, 23). They no longer see themselves as properly fitting their bodies (Gen 3:7-8). The woman's future progeny no longer properly fits her own body (Gen 3:16). The couple no longer properly fit each other (Gen 3:16). Though more ambiguous in nature, the humans have lost any hope of proper fit with these other creatures descending from the serpent (Gen 3:15), a shrewd serpent now mis-fitted to his food and reranked among the other animals (Gen 3:14). *Only in the curses do we find how thoroughly fitted together the whole scheme of Eden had been all along.*

Taken as a whole, the texts of Genesis 2–3 appear to interconnect *dirt* to *animal* to *orchard* to *woman* to *dirtling* to *serpent*, and ultimately to God. This theme of interconnection persists into the flood narrative, where the conflict resolution aims at bringing "relief" (*nakham*) from the curse of the dirt (see Gen 5:29; 8:21). God intensifies the connectedness of all living things with "the breath of life" (Gen 7:22) by constraining himself in his promises to Noah's progeny and all animals ("all living things"/*nephesh hayot*) in parallel to humans (Gen 9:9-11). God covenants with animals, not the trees or waters, and holds them accountable for murder.[5] Noah *and* his family *and* the dirt *and* the animals are bound up together in the flood narrative as entangled agents of creation, where God's actions focus on the problem of misfit to the dirt and the unnecessary violence it produced (see Gen 3:18; 4:12-14; 6:11; 8:21).

---

[5]In Gen 9:5, God requires a reckoning from animals and humans alike for killing humans. However, fruit trees later receive a reprieve from conquest violence specifically because they are not humans: "If you besiege a town for a long time, making war against it in order to take it, you must not destroy its trees by wielding an ax against them. Although you may take food from them, you must not cut them down. Are trees in the field human beings that they should come under siege from you?" (Deut 20:19 NRSVUE).

EXILE AS DISLOCATION

If they do not fit, then God must exile them. Genesis opens with fit and then mis-fit to the dirt from which the dirtling was taken. Exodus focuses on Israel's future fittedness to the barren land of Canaan. In this schema, exile will become a kind of mis-fit. From the garden at Eden to the Babylonian exile, a recurring theme of fit-but-potentially-dislocated lurks in the history of Israel. Speaking generally, the land of Canaan provided food sufficiently if justice was maintained. If not, exile pried Israel from a land to which they were no longer fitted.

But Israel's mis-fit is a process, beginning in the long-suffering of God and Israel's exploitation of their own vulnerable folks. Israel's sins against the resident alien, poor, widow, daughters, and more all polluted the land, as did Canaanite behaviors prior to Joshua's conquest (see Lev 18:24-25; Deut 9:4-5).

Leviticus employs the image of the land that "became unclean, so that . . . [it] vomited out its inhabitants" (Lev 18:25). The analogy suggests a long, progressive buildup of corruption that culminates in their expulsion. After Canaan is purged of Canaanites (some areas more than others), Leviticus proleptically reports that the land itself will enjoy its rest, its Sabbath from wickedness (Lev 26:34). The Canaanites progressively mis-fit themselves to their environment, and the Hebrew conquest dislocated a mis-fit people. Israel eventually followed suit.

When compared to Israel's peers, it is not unheard of to find some modest concern for the poor and marginalized in Egyptian and Mesopotamian literature. However, only in the Torah do we find a tradition where a god promises to uphold the same justice *against Israel* if she continually misfits herself to her land through the prybar of systemic injustice.[6] It is of interest that God threatens to intensify the curse of the dirt, corrupt Israel's fruitfulness, plague her with sickness, and take her away into captivity to punish an injustice—all forms of misfit and dislocation (see Lev 26:1-39; Deut 28:15-68).

Exiled, Israel was mis-fitted to the land that eventually vomited her out. This mis-fit must be accommodated in some way. In Jeremiah's epistle to the

---

[6]Jeremiah Unterman, *Justice for All: How the Jewish Bible Revolutionized Ethics* (Philadelphia: Jewish Publication Society/University of Nebraska Press, 2017), 41-84.

exiles, he both reinforces the longer exile—seventy, not two years—and encourages Israel with the Edenic language of blessing and curse. This time, it is used positively: build (*banah*), settle (*yashav*), plant gardens (*nataʿ gan*), eat fruit (*ʾakhal peri*), take wives (*lakhah ʾishah*), give (*natan*) your sons to marriage, and multiply (*ravah*). In two sentences (Jer 29:5-6), Jeremiah employs most of the key terms of fittedness in Genesis 2–3, all of them if we include Jeremiah 29:7-11.

Amos also uses much of the same language and concepts found in Jeremiah's epistle to reinforce that Israel's fit to Canaan was the divine goal. Over the course of two sentences (Amos 9:14-15), Amos describes God's intent to return Israel to Canaan, where they will build (*banah*) cities, settle (*yashav*) them, plant (*nataʿ*) vineyards, make gardens (*gan*), and eat fruit (*ʾakhal peri*). This condensed use of identical language with similar symmetry and context makes the point unmissable. Like in the garden at Eden, unethical behaviors—mistreatment of the marginalized (see Amos 2:6-7; 8:5-6)—will cause mis-fit to an otherwise amenable home in Canaan. An accommodated fit is made in Babylon (e.g., Jer 29), but the intended fit consists in return to Canaan (see Ezek 28:25-26; Amos 9:14-15; cf. Amos 5:11; Zeph 1:13).

Reversing the curses of Eden, the eschatological vision of Isaiah 65 uses the same language and context to reassure of the time when the fittedness of Israel and the gathered nations will be complete, durative, and secure:

> They shall *build* houses and *dwell* in them;
>   they shall plant vineyards and *eat their fruit*.
> They shall not *build* and another *dwell*;
>   they shall not plant and another *eat*. (Is 65:21-22 modified)

As it was in the beginning, humans will be metaphysically fitted to their habitat. The natural relationship between land, people, homes, fields, trees, and politics is depicted consistently, even if paradoxically, as fitting and rightful.

THE ORIGINS OF LANGUAGE AND FIT

Remember that my goal has been to follow the intellectual tradition set out in the biblical literature. The question at hand is: How do the biblical authors construct concepts such as fit to environment? So far, we have been surveying the obvious candidate: the making of humanity and situating them in the niche environs of Eden. Oppositely, we have seen the factors of mis-fit to those environs, even naturally hostile environs such as Canaan, turn on the morality of the community of Israel.

Another way to conceptualize fittedness can be found in an admittedly odd place. The tale of Babel in Genesis 10–11 may appear as a simple morality play. But its position and content show why it is central to the intellectual world of the Hebrews. It is a story located in a seminal position meant to explain why everything is the way it is today (i.e., it is part of Israel's primeval history).

Physical fittedness is not the only concept of environment addressed in Genesis. Remember those environmentalist and anti-environmentalist debates of Darwin's day? The question could be put to Genesis: Are features of humanity defined by the locale of the creature, or are they divinely created?

The story of Babel, and its prequel in the table of nations (Gen 10), suggests an environmentalist view of human language (i.e., that different languages emerge from humanity's fittedness to various habitats).

Research on human phonics and linguistics might back up Genesis on this point. That research has favored a hypothesis about the direct effects of location on language generation. In short, the physical terrain of a location forms the development of the language spoken in that terrain. This includes how a terrain-shaped language diffuses into dialects and how those dialects spread.

This research offers obvious explanations about the nexus of terrain and language. For example, when uncrossable rivers separate people groups, new dialects are created. More unusual findings have emerged as well. High altitude increases the likelihood of consonantal ejectives appearing in a language.[7] (An ejective is a harsh consonantal sound found in languages such as Georgian, Navajo, or Quechua, but not found in most Western European languages.) The higher up a population lives, the more likely they are to have some of these harsh consonant sounds in their language or dialect.

Heat also appears to distort consonantal sounds when spoken over distance. So, words with complex-sounding combinations of consonants were

---

[7]Caleb Everett, "Evidence for Direct Geographic Influences on Linguistic Sounds: The Case of Ejectives," *PLoS ONE* 8, no. 6 (2013): e65275; Stephen C. Levinsona et al., "Returning the Tables: Language Affects Spatial Reasoning," *Cognition* 84 (2002): 155-88.

found to be rarer in hotter climates. Consider the following example: "Geography can be a real problem for complicated consonant-rich sounds like 'spl' in 'splice' because of the series of high-frequency noises. In this case, there's a hiss, a sudden stop and then a pop. Where a simple, steady vowel sound like 'e' or 'a' can cut through thick foliage or the cacophony of wildlife, these consonant-heavy sounds tend to get scrambled."[8] As rivers and mountains carve up migrations and create pockets of settled people groups, the topography then fashions both the spread—or not—of dialects as the language adapts to new environs. (And in our present language development, what topography once shaped, our digital media landscape now governs.[9]) This explanation of linguistic shape and spread is potent enough that one study proposed that language diversity could be explained best by only two of the fourteen factors they considered: river density and topographical jaggedness.[10]

If the environmentalist views are reflected in the biblical literature, then we should expect to find discussions of language diversity and language propagation that focus mostly on location as an explanation for diversity. In this conceptual world, language and habitat should exhibit a socio-geographic view of linguistic fit to environment.

This simple connection flags up a story that Genesis particularly wants to tell, but with a surprising twist. The account of Babel in Genesis 11 seems intent on describing the origins of language diversity. Reading the renowned

---

[8]Reporting on Maddieson's work: Angus Chen, "Did the Language You Speak Evolve Because of the Heat?," National Public Radio, www.npr.org/sections/health-shots/2015/11/06/454853229/did-the-language-you-speak-evolve-because-of-the-heat/. Ian Maddison writes, "Specifically, environments in which higher frequencies are less faithfully transmitted (such as denser vegetation or higher ambient temperatures) may favor greater use of sounds characterized by lower frequencies [i.e., vowels]. Such languages are viewed as 'more sonorous.'" Maddieson, "Human Spoken Language Diversity and the Acoustic Adaptation Hypothesis," *The Journal of the Acoustical Society of America* 138, no. 1838 (2015).

[9]James Burridge, "Spatial Evolution of Human Dialects," *Physical Review X* 7, no. 031008 (2017).

[10]Jacob Bock Axelsen and Susanna Manrubia, "River Density and Landscape Roughness Are Universal Determinants of Linguistic Diversity," *Proceedings of the Royal Society B* 281 no. 20133029, June 7, 2014. "The significant relation between river systems and LD [linguistic diversity] here uncovered may hopefully open new ways of looking at how languages, and thus human cultures, arise, interact and can be preserved. If linguistic recombination is behind the proliferation of languages along fluvial networks, linguistic phylogenies in those regions may appear as complex ensembles of networked languages, with a strong component of horizontal transmission that may eventually call for an analysis of the parallelisms between linguistic and biological evolution beyond the analogy."

story in the context of the prior genealogy may cause a different portrait of diversity to emerge.

What, exactly, diversified language at Babel? From Augustine to Dante, many have read Genesis 1–11 as a story that begins with a pristine Edenic language ultimately confused by God. Of course, that language was presumed to be Hebrew or some other universal language that was diversified when God scattered all humans from Babel.[11] While an appreciable interpretation, this reading might miss the literary context of the Babel story. That context explicitly tells us *who* went into Shinar and built the Mesopotamian cities—namely, Nimrod, the descendant of Ham (Gen 10:9-10). Among those cities that Nimrod built, we find Babel (Gen 10:6, 8-10):

> The descendants of *Ham*:
> > *Cush*, Egypt, Put, and Canaan.
> > *Cush* became the father of *Nimrod* . . .
> > The beginning of [*Nimrod's*] kingdom was *Babel*, Erech, and Accad,
> > all of them in the *land of Shinar*. (my translation)

The table of nations in Genesis 10 sticks out as a unique genealogy in the Hebrew Bible. It is unique because the names of Ham's descendants are the names of locations, both cities and empires renowned to Israel. Like some of Cain's descendants, these descendants of Japheth, Ham, and Shem are notably enumerated not just as founders of industries but as political entities. They were builders of cities, empires, and regions. In short, the enigmatic genealogy of Genesis 10 tells the reader about geopolitical realities as much as family lineages.

Further complicating the genealogy, Genesis's narrator not-so-subtly punctuates their city-building genealogies three times with the explanation for the diversity of languages: "These are the sons of *x* by their clans, their languages, their lands [*be'artsotam*], and their nations [*goyim*]" (see Gen 10:5, 20, 31). The genealogy opens and closes with humanity's postflood dispersion, but note what creates the lines of delineation: their kin, their land, their politics, *and their languages* (Gen 10:1, 32).

Genesis 10 renders Noah's genealogy topographically. In the ancient mind, this genealogy creates an image of the mountains, deserts, and valleys

---

[11]Livingstone, *Adam's Ancestors*, 49-50.

on which Noah's descendants were spilled out. It was the geographies into which they fit and the clans to which they belonged that shaped their diverse political entities and languages.

I suggest that we read Genesis 11 with the environmentalist views that Genesis 10 offers us about languages: that different languages are a product of diverse clans in different locales.[12]

The story immediately following that geopolitical/linguistic genealogy of Genesis 10 then focuses on the construction of one particular city: Babel, in the land of Shinar. The opening of the story might distract those of us who live in the era of space pioneering. Today, when we read, "Now all the earth [ha'arets] spoke as one and had one [set of] words" (Gen 11:1 my translation), surely we picture a globe hanging in space as "the whole earth." However, it could equally read, "It came to be, *all the land* spoke as one and had one [set of] words."

What land? The land in which the people came east, which includes Shinar (Gen 11:2). What people? The people Genesis 10 just informed us about, Nimrod's descendants. What is so special about Ham's descendants from Nimrod? Genesis notes only that they built cities in Mesopotamia and explicitly "the land [ha'arets] of Shinar." As it would not make sense to translate "the earth of Shinar" in this context, it might also be misleading to translate "the whole earth" in Genesis 11:1. The same term "land" ('erets) used in Genesis 11:1 is sometimes translated as "the earth," except that the context might require us to say "the whole land."

What are the names of those cities in the plains of Shinar? "Babel, Erech, Accad, and Calneh, in the *land* of Shinar [he'erets shinar]" (Gen 10:10).

So, the unique city-building genealogy of Genesis 10 tells us about a city called Babel, built by Nimrod's descendants, and then the next story depicts the building of Babel in the "land" where everyone spoke the same language (see Gen 10:10; 11:2). It is difficult to see how the Babel story is about all humans on the earth when we have been specifically told that it was just a subset of Ham's progeny who headed east to build Babel.

---

[12]But does Gen 9:19 not say that all the people of the earth were dispersed from the three sons of Noah? And God not disperse the people of Babel? Given this statement, followed by the genealogy that tells us which descendants explicitly went into Shinar to build Babel, and then the Babel story, we cannot unhear what we know about Babel if we read in the literary order.

Why is this important? Not only does the "whole earth" translation unnecessarily skew our minds to believe it means "every human alive," but we ignore the literary context that the genealogy means to explain the mechanism of language diversity. Prior to the building of Babel, Genesis 10 explains the postflood human languages that were formed by different lineages, locations, and leaderships.

The story of Babel focuses on the technology and ambitions of Nimrod's descendants to buffer against scarcity, as we saw in previous chapters. They sought physical and agricultural security, not linguistic unity. Linguistic unity and diversity function in the narrative as both the setting and punishment's outcome, respectively.

Their scattering "over the face of the land [*ha'arets*]"—repeated twice in this passage—might at first seem to result from the divine confusion of their languages. This would be the anti-environmentalist view of Babel. God made languages for places, suggesting that different languages would cause peoples to separate into different places (Gen 11:8-9).

However, given the depiction of language diversity in Noah's genealogy, it should read the opposite direction. This monolingual subset of Ham's descendants was dispersed, and the new locations and division of peoples created new languages. Emerging languages were fitted to the new places to which they were scattered. This would be an environmentalist view.

So which is it? The narration of the story is terse and to the point. God seeks to "confuse their language" (Gen 11:7), yet the only action God takes is to disperse them, which is precisely their only stated fear. Here is the short version of the story:

> **People of Babel:** *Come, let us* build . . . *lest we be dispersed over the face of all the land.* (Gen 11:4 modified)

> **God:** *Come, let us* go down and there confuse their language. (Gen 11:7)

> **Narrator:** So Yahweh *dispersed them from there over the face of all the land.* (Gen 11:8 modified)

> **Narrator:** Yahweh confused the language of all the land. (Gen 11:9 modified)

> **Narrator:** And from there, Yahweh *dispersed them over the face of all the land.* (Gen 11:9 modified)

Genesis 11, like the previous genealogy, might be attempting to show how languages emerge topographically. Genesis asks us to imagine the mountains, deserts, and valleys on which God scattered Babel's inhabitants. It was the geography into which they fell that shaped their political entities and languages. God says "come, let us . . . confuse their language," but God only acts to disperse them. Scattering is confusion because different locations create different languages. Even if they eventually found each other and regrouped, their new cultures and languages would keep them from uniting in obstinacy to God's creational blessings to "fill the land."

Taken together, Genesis 10 and Genesis 11 form a loose argument for the diversity of language (and culture) according to kin and terrain. Topography shapes language, possibly even in Genesis's only account of the origins of languages.

Of course, Babel is not the final account of linguistic diversity. The book of Acts portrays a remarkable reversal of Babel on Pentecost. Jews gathered into Jerusalem from diverse regions. The narrator presents a similar view of language and location to Genesis. The story notes that the Jews spoke languages according to those regions from which they came, and they could all hear Galilean men praising God *in their own languages* (Acts 2:5-12).

Notice that the characters and narrator presume that the different regions best explain their language diversity in Acts 2. These diaspora Jews who had come into Jerusalem for Shavuot (Pentecost) "were amazed and astonished, saying, 'Are not all these who are speaking Galileans?'" (Acts 2:7). Their perplexity at the ability of these men appears to stem from the improbability of a Galilean knowing their regional languages. Presumably, Galileans are not the kind of Jews who would or could travel to those faraway places, nor would they be the kind who could learn those exotic languages at home.

One also wonders whether the author of Acts had Ham's genealogy in mind when listing all the regions from which the Pentecost Jews had come. If one looks at a map of the descendants of Ham (Gen 10:6-20) and overlays it with a map of the regions of the Pentecost travelers (Acts 2:9-11), one might reasonably conclude that these are not two stories connected merely by the theme of language, but they are also connected by location.[13]

---

[13]Gilbert notes that the comparison between the table of nations in Gen 10 and Acts 2 never fully works out. I wonder whether restricting it to Ham's genealogy and recognizing some of the

Language and evolution. Among scholars of acoustics, linguistics, and anthropology, the evolution of language requires a coherent causal chain that can explain speech from hominin to psalmist. Humanity fits to habitat, even linguistically. Habitat fashions language.

But linguistic fittedness is not the kind of "fit to environs" that Darwin had in mind in the evolutionary struggle for life. So why raise the matter here? My goal has been to demonstrate that the same pressures that fund evolutionary theory are *conceptually present* and discussed as relevant to the arguments about the nature of reality that Genesis 1–11 wants to make.

By showing how fittedness features prominently in humanity's relation to all aspects of their survival and flourishing in Scripture, I hope to have illustrated the prominence of the theme across a portion of the biblical literature. By demonstrating fittedness outside the realm of biological fit, I hope to have begun a case for a widely employable notion of fittedness as a positive feature, as it is in Darwin's thought, but with a divine agenda to it.

REFITTING US TO NEW CREATION

Finally, there is a theme of fittedness that runs through the eschatological discourses of the New Testament worth considering. The striking revelation of Scripture is that the "Israel of God". must be made to fit the new heavens and new earth. They are presently incompatible with it.

Briefly, Jesus bases his own understanding of human affairs on a creational (or natural) paradigm. This fitting-together and tearing-apart paradigm mirrors the concepts of those who will be fit for the coming kingdom and those who will not.

Jesus' rebuke of divorce conceptually highlights the divide. When Pharisees approach him about the permissibility of divorce in the Torah, Jesus immediately refocuses the discussion on creation looking forward. In creation, God fits together two sexually differentiated beings in their blessed and created state. Jesus goes on to highlight that divorce was an accommodation for the cursed state of human "hardness of heart." He then advocates

---

uniquely distinct entries (e.g., Caphtorim/Crete, Lehabim/Cyrene, Egypt/Egypt, Persia, Mesopotamia, and Arabia) offers enough support to say some connection is being made by Acts 2. Gary Gilbert, "The List of Nations in Acts 2: Roman Propaganda and the Lukan Response," *Journal of Biblical Literature* 121, no. 3 (Autumn 2002): 497-529.

that "what God has joined together, let no human tear apart" (Mt 19:6; Mk 10:9 my translation). He reminds them, "From the beginning it was not so" (Mt 19:8).

This divine fitting together and separating based on the heart—an organ of human reason and deliberation—is a deep structural theme carried to the end of the biblical story line. Jesus' attention to the created fittedness of two humans by way of the bride-groom relation does not end in the Gospels.

Jesus' teaching also included dichotomies of those who would fit into the coming kingdom and those who would not. Who are those not fit for the kingdom habitat? Examples include teachers who relax the Torah (Mt 5:17-20), those who do not receive the kingdom like a child (Mk 10:13-16), those who merely cavort with the Way of Jesus (Lk 13:22-30), those who are not "born of water and spirit" (Jn 3:3-5), and more.

Even Revelation's descriptions of the eschaton highlight both the renewal of creation (i.e., the new heavens and new earth) and the various ways in which we will be fitted to it. We now recognize all the themes of provision and fit. The renewal of the heavens and earth appears to mean the uniting together of the two realms, as Revelation uses the analogy of joining a bride and groom (Rev 21:2). In this reorientation of creation, there is no death. Resurrection returns the redeemed to their created estate (Rev 21:4). Those not worthy of the "age of resurrection" (e.g., Lk 20:34-40) are described as mis-fitted to the full kingdom of God (e.g., false, detestable, and unclean persons, Rev 21:27). In an unmissable throwback to the creation narrative and the origins of this renewed cosmos, food bears from a twelve-month "tree of life" constantly fed by "water of life" (Rev 22:1-2). All of this is politically administered and orchestrated by God himself.

Metaphysically disoriented humans and creation are reoriented to fit together with creation and with God. As Paul puts it in his letter to the Ephesians, it is God's "plan for the fullness of time, to unite all things in [Jesus], things in heaven and things on earth" (Eph 1:10). What has been torn apart will be fitted back together.

Even in the repetitious referrals to that renewal of the heavens and earth, it is through a glass darkly that we glimpse the new renaturalized world to come, to which I will return in the final chapter.

STICKING POINTS

What are the agreements and conflicts between the conceptual schemes of fittedness in *some* evolutionary science and what we find across Scripture? First, the environmentalist's idea of generations shaped by habitat demands some account of fittedness among creature, culture, and environment. These factors of fit figure into any calculation about the genetic shape of any given creature and the trajectory of evolution more broadly. The environmentalist view also means that primacy is placed on habitat in thinking about the development of emerging features (e.g., new languages adapted to new topographies). At least one biblical scheme of environmentalism appears in the diversity of language in Genesis 10–11, possibly persisting into Pentecost (Acts 2).

Second, mis-fit to environment occurs by moral violation in the biblical texts. Unlike many accounts in the evolutionary sciences, fit to a divinely constructed environment entails moral and communal obligations. Nature, even when denatured, is not morally neutral for humans or animals. This requires humans and a God who attend to their relationships and obligations to creation and creatures alike. It is specifically not a course of events guided by so-called natural laws.

Paul connects these two realms in many places, but his letter to the Colossians (Col 1:16-20) puts the matter plainly. He begins by placing Jesus in the creation and then as the present metaphysical sustainer of the physical cosmos. Paul depicts Jesus' totalizing authority and control of the cosmos through the incessant use of the term *all* six times in a few sentences. He then connects the metaphysical rule of Jesus over the cosmos to Jesus' goal of reconciling us to peace (i.e., fit to the kingdom) through resurrection.

Biblical accounts of maladaptation to habitat (e.g., vomited out of the land, scattered, or destroyed) and habitat accommodation (e.g., Canaan as land of "fat and honey") assume the current metaphysical disorientation of the cosmos. There is no place or creature with a fully conducive fittedness to the current environment. Such a fit is only anticipated in the new heavens and new earth. Biblical thinkers put fittedness with "the land" in parallel with fittedness in the new heavens/earth, which requires being properly related to God (ethics), to earth (metaphysics), and to others (personal/relational). I cannot yet imagine a form of evolutionary explanation that would have reason to entertain these moral and theological dimensions of fittedness, but there is no logical need to exclude them.

PART FOUR

GENERATION

CHAPTER TEN

EVOLUTIONARY SEX AND EXTINCTION

Monogamous heterosexual love is probably one of the most difficult, complex and demanding of human relationships.

THOUGH REGARDED AS NORMAL in much of the world, the ideal of lifelong sexual monogamy has some explaining to do. Evolutionarily speaking, it does not make much sense depending on how one constructs one's ideals about evolutionary goals.

As it turns out, a father who will mate, play, and stay with his offspring benefits the whole family unit beyond an animal's ability to calculate those benefits. So what gave rise to monogamy? While we are asking questions, what gave rise to concerns about sexual generation (i.e., my descendants) beyond reproduction (i.e., my children)? Is the former more likely to benefit communities over the latter?

Because evolutionary biologists presume that forcible sex has been the primary mode of reproduction among many animals, is there any principled ground for demanding that consensual sex should be the norm? Hopefully, we all want two consenting adults to be the norm, but is that norm rooted

---

The quote in this epigraph is widely attributed to Margaret Mead in books and across the internet, but I could not find an original source.

in anything but the ever-fluctuating biases of cultures and tradition? It does not take long to figure out that the motivations and rewards of sex are mysterious, even to those having it.

George Michael once sang, "I want your sex." On its own, it does not make much sense. But somehow, we all know what he means. By saying it strangely, Michael points out that sex is more than the physical act. Philosopher Michel Foucault once famously wrote that "sex is worth dying for."[1] It is another odd phrase meant to highlight all the things that surround sex: the desire to have it, beauty ideals that drive body modification, hazardous behaviors because of desires to copulate or couple, libertine use of sexualized bodies to advertise harmful products, and a host of caught and taught ways of acting in the world. Because of our inability to fully understand our relationship to sex, we will careen forward wildly in all things sex. Evolutionary sex, if I may call it that, also puts its own spin on death and reproduction.

In our final foray into natural selection, we will touch on sensitive grounds: sexual reproduction, sexual assault, and generation. Here, the goals of sex will come to the fore as the most ethically troublesome aspect of natural selection. The kind of sexual partnering in mind entails these goals, whether it be reproduction (making babies), the urge to merely copulate or orgasm, or generational intent that looks beyond the immediate generation or sex act. This looking beyond appears to be an intention mostly limited to humans. In this chapter and the next, we explore the most bizarre, risky, and uniquely human sexual behavior that needs a good explanation: partnering and parenting beyond the mere act of sex itself (i.e., monogamy).

Here in the cauldron of individual anxieties about extinction and the restraint of sexual desires we find the most dramatic differences between the origin stories in evolutionary biology and the biblical literature. Whether we are an atheist or believer, we have largely sided with the biblical concept of sex. But why?

The biblical accounts certainly contain unrestrained reproduction, generation, and sexual assault (and that is just in Genesis!). Yet, the overall

---

[1]Michel Foucault, *The History of Sexuality, Vol 1: An Introduction*, trans. Robert Hurley (New York: Pantheon Books, 1978), 156.

picture of sex that emerges across the biblical texts favors a kind of sexual partnering where a couple stays and plays with their children. Moreover, and this is strangely important in Israel, the couple must practice the theology least likely to cause them to murder or prostitute their children in rituals to other gods. Real people, real problems.

In researching this topic, I was a bit scandalized to realize that sex is not even biologically necessary to reproduction. I suppose I knew this from high school biology, but it just had not registered as that remarkable of a fact: asexual reproduction—like cells do—could have been the norm.

This world could have been entirely sexless! Now we are ready to start thinking about how radical sex really is to creation. Imagine a world of hominins where sex differences and the desire for sex, coupling, relationship, and more were stripped out of the system. What would that be like? I have trouble even imagining it, if I am honest.

If sex could have evolved from asexual reproduction, then against every impulse of the biblical creation accounts, sex is not a necessary feature of creation. In Genesis 1–2, sexual difference is the route to flourishing, and sexual difference gets set as the cornerstone of creation: "in the image of God he made them, male and female he created them" (Gen 1:27 modified). Only one reason for sexual differentiation exists: sexual reproduction. Most, if not all, of the other distinguishing traits of males and females flow from their reproductive differences. To make the male-female distinction part of the *imago Dei* means that sexual reproduction represents the *imago Dei* as well. In evolutionary terms, the image of God reflects a more controversial ideology than we might first imagine.

WE DON'T EVEN NEED SEX?

Given the ubiquity of sex, it is easy to forget that
it is not necessary for reproduction.

NICK COLEGRAVE, "THE EVOLUTIONARY SUCCESS OF SEX"

Not only is sex not necessary, but some biologists also believe that asexual reproduction was foundational for the biological world. Asexual reproduction refers to reproduction without male and female gametes involved.

Cell division, spore budding, and fragmentation of starfish or flatworms are all examples of asexual reproduction.

Sexual reproduction is presumed to have emerged from asexual propagation, contributing a complicating impact on diversity in the story of so-called natural history. This helps to explain why the very notion of sexual reproduction confused Darwin: "We do not even in the least know the final cause of sexuality; why new beings should be produced by the union of the two sexual elements, instead of by a process of parthenogenesis [asexual reproduction]. . . . The whole subject is as yet hidden in darkness."[2]

Sexual reproduction confuses the evolutionary story because it is exceptionally onerous for most species, from birth to death. Onerous equals death and extinction in most evolutionary accounts. As we will see in the next chapter, entire species arrange their lives *and deaths* around sex, not just humans. From menstruation to physical coupling, human reproduction is a huge drain on resources. As mentioned in previous chapters, females' preferences in partners can contribute to precarious genetic situations for the males of the species. Darwin noticed and puzzled over why females will choose and propagate genetic features that appeal to them in mating rituals. Yet these preferences that get selected can also make males less fitted to their environment, reducing their individual survival. Presumably, this could improve the gene pool. For example, any male bird that can survive to mating season, despite being brightly colored or burdened by ornamental features, would certainly be a more favorable sexual mate.

What does sexual reproduction offer in natural selection? The evolutionary advantage of sexed differences was assumed to be the genetic variability created by male and female genetic recombination. More genetic variety was thought to produce more chances for fittedness and propagation, so the thinking went. However, population genetics has shown that the quality of fittedness created by sexual reproduction could instead reduce an organism's chances of fit to habitat.[3] In other words, sexual partnerships do not necessarily offer genetically fitting advantages in natural selection.

---

[2]Charles R. Darwin, "On the Two Forms, or Dimorphic Condition, in the Species of Primula, and on Their Remarkable Sexual Relations," *Journal of the Proceedings of the Linnean Society of London* 6 (November 1861): 77-96.

[3]"It is often rather glibly assumed that there is an obvious advantage to sex from generating increased variability; however, population genetic models show that sex and recombination can actually

Despite the presumption of sex's origins arising out of asexual reproduction, its arrival supposes a less sexy but theologically significant point: if
sex evolved from asexual reproduction, then there might have been a time
when sex was not significant. Because the biblical authors think in terms of
Edenic fruitfulness and generation, a basic value gap emerges. For the biblical authors, sexual difference in animals and humans constituted creation
as good and divinely intended, not an onerous process gone awry.

THE GREATEST NUMBERS OF OFFSPRING

Amongst many animals, sexual selection will have given its
aid to ordinary selection by assuring to the most vigorous and
best adapted males the greatest number of offspring.
CHARLES DARWIN, *THE ORIGIN OF SPECIES*

To state the obvious: sexual reproduction centers on the desire, skill, and
success of sexual copulation between males and females of reproductive age
and season. It is a physical act that we might associate with mating rituals,
but animals with such rituals do not represent all sexual reproduction. In
the history of living organisms, most scientists seem to think of these rituals
as a later development—where females select males with whom they have
"consensual" sexual encounters for the sake of reproduction. What came
before such rituals?

Debates simmer in evolutionary biology and psychology regarding *Homo
sapiens*'s sexual encounters. Specifically, how did hominins navigate sex to
the point where consensual monogamy became globally normal for large
swaths of humans? Setting aside the difficulties of maintaining monogamy
and love between humans, *sexual pairing* (i.e., Ms. Right) as opposed to
pairing for the purpose of sex (i.e., Ms. Right Now) poses difficulties at every
angle of explanation.

I begin this exploration of sexual propagation with the most problematic
aspects of sexual practices within evolutionary stories and end with Darwin's
tree of life, which is the sexual-propagation chart par excellence.[4]

---

reduce variability in fitness, given certain types of interactions in fitness effects among genes." Brian
Charlesworth, "The Evolutionary Biology of Sex," *Current Biology*, September 5, 2006, R693-R695.
[4]Don't get too excited; it's just a tree chart.

Cooperative sex. Let's start with the more palatable matters of sex among animals. Regarding cooperative sexual pairing, the female's enculturated preferences will factor in her selection of sexual partners. This creates a wrinkle for natural selection, as previously discussed in chapter eight, because the female preferences might not be genetically advantageous features. Philosopher Mary Midgley waxes eloquent here about the male pheasant's mating feathers:

> Who, however, designed these wonderful feathers? Certainly not a human. If Darwin is right, the artists here seem to have been the birds themselves, which means . . . chiefly their wives. . . . Darwin was suggesting that the wishes of hen-pheasants—their inner thoughts and feelings as they watched their various suitors—had affected, and had finally determined, the design of later generations. . . . The difficulty is quite general.[5]

Preference drives selection. But it is not necessarily mere preference. Preferences could conceivably drive sexual selection in a way that aims the female's attention at genetic features that the female could not even articulate. Her preferences may have roots in genetics, in which she only plays a part. If this is the case, genetic determination might still drive selection even when cultural preferences have significant influence.

Presumably, some combination of culture, genetics, and epigenetics exists within mating rituals. I have already discussed the difficulty of distinguishing cultural preferences from genetically rooted mate selection in chapter eight. But even the mixture itself hints at a problem with the once-taken-for-granted idea that natural selection alone will propagate creatures that are most genetically fit to the environment. In humans, we like to grant a degree of freedom in our choices, even sexual choices, that can be coerced but not determined by factors internal and external to us. This will factor directly into the discussion of rape below. The difficulty is, indeed, quite general.

Regarding the practice of merely *pairing for sex* among most mammals, copulation is the *end* of that relationship, in both senses of *end*. This is sometimes true among humans too. However, research on male parental involvement across mammals shows that it might ultimately be

---

[5]Mary Midgley, *Are You an Illusion?* (New York: Routledge, 2014), 74-75.

advantageous to monogamously pair beyond copulation.[6] Male parenting appears to stem from monogamy, but no one yet knows why monogamy emerged.[7] It's unclear how the males of any given species could ever become aware of such a long-term benefit of pairing monogamously, aware to the point that it factored in their parenting behaviors over the life of their children.

Considering the origins of monogamy, one research team suggests: "Perhaps monogamy did not evolve in the genetic sense at all, but rather in a cultural sense. . . . Humans and their anscestors [*sic*] are too sexually dimorphic in size to be considered naturally monogamous."[8] By "too sexually dimorphic" they mean that unlike many other species, a human male is both structurally larger and can impregnate many women constantly and year round. Conversely, any particular woman can only be impregnated during one brief span per month. This difference in reproductive ability alone would seem to argue against humans being "naturally monogamous" because the stronger and more sexually prolific males do not appear to have any incentive to stay and parent with one particular female. Or, as Matt Gage suggests, "Sperm-producing males tend to have huge reproductive potential and are limited [only] by the numbers of mates they can fertilize; selection therefore acts on males to mate multiply and realise their evolutionary potential."[9]

Uncooperative sex.

That the males of all mammals eagerly pursue the females

is notorious to every one. . . . The female, on the other hand,

with the rarest exceptions, is less eager than the male.

CHARLES DARWIN, THE DESCENT OF MAN, AND IN RELATION TO SEX

---

[6]"It is often in the best interest of a male to forgo pursuing multiple mating opportunities and instead achieve high paternity certainty with a single partner." Ryan Schacht and Adrian V. Bell, "The Evolution of Monogamy in Response to Partner Scarcity," *Nature/Scientific Reports* 6, no. 32472, September 7, 2016, doi:10.1038/srep32472.

[7]"Paternal care is more likely a consequence of monogamy—an evolutionary afterthought with benefits—than the key to its existence." Though researchers ruled out male infanticide of rival babies as a cause of monogamy, they were less confident in ruling in the reasons for it. Frans B. M. de Waal and Sergey Gavrilets, "Monogamy with a Purpose," *Proceedings of the National Academy of Science*, September 17, 2013, 15167-68.

[8]De Waal and Gavrilets, "Monogamy with a Purpose."

[9]Matt Gage, "Evolution: Sexual Arms Race," *Current Biology*, May 25, 2004, R378-R380.

The enigmas of cooperative and monogamous sex among humans also lead to the aspect most difficult to speak of with remove: uncooperative sex as propagation. To be clear, "uncooperative sex" would be called sexual assault or rape, in human contexts today.

The sexual eagerness of males that Darwin identifies above appears to be tacitly accepted among biologists. Therefore, their assumption of monogamy's emergence from a world of male promiscuity leads to the dark reality also presumed within most evolutionary origin stories: sexual assault essentially fueled some or most of hominin reproduction. Not only can males inseminate females year round, but those who have the libido to do so also lack reasons for staying, raising, and playing with their offspring, as it were. Additionally, males often have the physical means to propagate uncooperatively, which is too-politely termed "forced copulation" across the evolutionary psychology literature.

Though some are reasonably cautious to call it *rape* in the animal kingdom, it is difficult to conceive of it otherwise among some species (and we would certainly consider it sexual assault if it were humans).

Regarding male ducks, which are generally larger than the females: the authors of a particularly descriptive essay describe "sexual conflict" in cases where the desire to mate differs between the male and female. One researcher paints the gruesome picture of sexual conflict this way:

> Males constantly harass females into mating at a rate that benefits selfish male interests, but is harmful to females. Females are sometimes drowned during ardent male mating attempts. This is one overt example of a mating pattern that exhibits conflict, but the reproductive costs for females as a direct result of selfish male interests can span a broad continuum, from death and injury, through parasite transmission and infanticide, to increased predation risk and reproductive time wasting.[10]

Regrettably, this behavior does not necessarily diminish the profitability of natural selection. Researchers found that "sexual conflict can promote speciation," which increases the odds of a genetic line propagating. While increased sexual conflict also seems to make females more selective about mates, it does not protect them from larger males bent on forced copulation.

---

[10]Gage, "Evolution: Sexual Arms Race."

Across the animal kingdom, most males are smaller than females. This means that forced copulation is a problem specific to certain animals where the males are larger.

Uncooperative sexual encounters are not limited to animals such as ducks, guppies, and the like, but also involve bottlenose dolphins, chimpanzees, orangutans, and humans.[11] To be fair, sexual coercion and forced copulation appear to be nearly universal among things that move on earth: "Examples of sexual coercion in different forms have been documented in all classes of vertebrates and in many of the better known invertebrates."[12]

In 2000, Randy Thornhill and Craig T. Palmer's book *A Natural History of Rape: Biological Bases of Sexual Coercion* incited debate about whether forced copulation was an adaptive strategy hard-coded into males and passed down genetically to human males today. Many scholars forcefully responded to Thornhill and Palmer's claims. Some responses focused on questions of methodology in evolutionary psychology, but the central question remained the same: Are males genetically disposed to rape—to propagate their genes by all means and, in all opportunities, when they are physically capable of doing so? Or, more bluntly, is forced copulation a biologically necessary part of the so-called natural history of sexually differentiated life?

Some have suggested that sexual coercion is part of evolutionary competition and success. It shapes an animal's body tactically for such sexual encounters: "This sexual antagonism between the sexes within the majority of mating patterns therefore generates an evolutionary 'arms race' in which males evolve adaptations that benefit their own reproductive interests, and females then evolve counter-adaptations."[13] Basically, when one sex is

---

[11]E. Scott, J. Mann, J. Watson-Capps, B. Sargeant, and R. Connor, "Aggression in Bottlenose Dolphins: Evidence for Sexual Coercion, Male-Male Competition, and Female Tolerance Through Analysis of Tooth-Rake Marks and Behaviour," *Behaviour* 142 (2005): 21-44; C. D. Knott and S. M. Kahlenberg, "Orangutans in Perspective: Forced Copulations and Female Mating Resistance," in *Primates in Perspective*, ed. C. J. Campbell et al. (Oxford: Oxford University Press, 2007), 290-305.

[12]Jana J. Watson-Capps, "Evolution of Sexual Coercion with Respect to Sexual Selection and Sexual Conflict Theory," in *Sexual Coercion in Primates and Humans: An Evolutionary Perspective on Male Aggression Against Females*, ed. Martin N. Muller and Richard W. Wrangham (Cambridge, MA: Harvard University Press, 2009), 28.

[13]Gage, "Evolution: Sexual Arms Race."

stronger, male and female bodies will adapt to resist or protect from the sexual act.

It's difficult to know what to make of such interpretations of evolutionary morphology and biology. But the question lingers for all who want to affirm theistic evolution: How does forced copulation fit into the divine formation of animals, including humans?

Monogamy. Monogamy is not evolutionarily advantageous. It does not make sense. It is an arrangement that we see in Eden, before anything goes wrong, that eventually reemerges in Judaism. I would argue that all polyg-ynous relationships depicted in Scripture demonstrate the universally neg-ative outcomes of polygyny (i.e., multiple wives): strife, exploitation, deceit, sexual misbehavior, sexual assault, and attempted murder between and within polygynous family systems (e.g., Cain, Abraham, Hagar, Jacob, Reuben, David, and so on). It is no wonder that Jesus points back exclusively to Genesis 2, not Genesis 12–36, for his source for reasoning about marital relations (Mt 19:8-9).

Even more puzzling for this whole discussion within evolutionary psychology: How did monogamy emerge in light of brute sociological factors that weigh so heavily against it in sexual reproduction? Maybe, as Ryan Schacht and Adrian Bell suggest, the sheer ratios of males to females produced the oddity of monogamy. Congealed with other factors to create the pot within which monogamy could simmer, "an abundance of men is associated with higher rates of relationship commitment, monogamy, lower reproductive skew among males, and less promiscuity in both sexes."[14] Maybe the male-to-female ratio itself made monogamy a thing.

The monogamy hypothesis attempts to resolve the puzzle of this odd be-havior by supposing that males had compelling reasons to exclusively pair with females and vice versa, despite not knowing the precise reasons for these pairings. When there are lots of guys around, those guys are more likely to see the need to compete for sex. Such explanations border on what Stephen J. Gould once referred to as a "just-so story," or an etiological folk-tale.[15] Again, the realities of propagation-by-forced-copulation sit right

---

[14]Schacht and Bell, "Evolution of Monogamy."

[15]Stephen Jay Gould first discussed this idea throughout his essay "The Return of Hopeful Monsters," *Natural History* 87, no. 1 (1978): 519-30.

under the surface of these origin stories: "The monogamy hypothesis suggests that pairbonding and male care preceded the emergence of cooperative breeding in our lineage." In other words, prior to males staying and playing with their offspring, there was a time when monogamous cooperative sex was either nonexistent or not normal.

When we consider sexual propagation on a continuum, evolutionary advantages pile up on one side. If we placed the least efficient cooperative sexual encounters of lifelong monogamy at one end, the other end would surely feature promiscuous forced copulations by males as advantageous to propagation.

The matter of monogamy keeps resurfacing here because it is a bizarre human behavior among living creatures. Humans are not the only animals to practice monogamy-like pairing, but evolutionary psychologists agree that the reasons fueling human and animal monogamy are probably not the same. Though never practiced uniformly across the human species, many sectors of global human history have held up monogamy as an ideal. In other words, monogamy is not an African, Asian, or North American phenomenon.

It holds a special place among sexual propagation practices because it is the least efficient way to spread one's genetic line and the most precarious practice in a survival or subsistence scenario.

Why is monogamy so precarious? It's rather inefficient and risky, mainly because monogamy relies too much on individual preference. The idea is this: females begin to select partners according to their preferences, and this creates a disruption in genetic variance, which otherwise would trade on the whims of chance mutations. But a female's preferences might possibly be rooted in evolutionarily advantageous features that tacitly factor into her selection. She might think it is a preference when it is actually a cultural-genetic pathway she is following.

Evolutionarily speaking, a male who looks good *right now* to a female might ultimately be "good-looking" based on genetic clues that a female could not articulate but are discernible to her nonetheless. If we were to speak only from the perspective of successful reproduction, a female's genetic instincts about sexual mates could tap into apt survival features more than her enculturated partner-shopping preferences ever could.

That's the best just-so story we can tell about monogamy's success in selection.

Conversely, the bizarre practice of lifelong monogamous coupling risks the future in several ways. The male-at-present (Mr. Right Now) could eventually become dangerous to her or her children. Or he could become deformed, cancerous, dead, or just a dud in the future. Indeed, humans wrestle with this same question: On what basis, then, should someone select a spouse with which to reproduce (i.e., Mr./Ms. Right) when all we have to go on is Mr./Ms. Right Now?

The most extreme form of this kind of coupling can be seen in China's one-child policy. An exclusively coupled family has one child to carry their name and genes forward into the population. That child will then marry one partner and produce one child, and so on. The precariousness of this social arrangement is being felt today, in cases where that one child grows into adulthood and dies before ever producing a child. With their childbearing years behind them, the monogamous couple cannot renew their family as Eve did with Seth when confronted with a similar scenario ("God has appointed for me another offspring instead of Abel, for Cain killed him," Gen 4:25). Their names, traditions, and genetics die with their one child. Such couples have been tragically called orphan parents. This kind of monogamy makes little bare sense in evolutionary models.

DARWIN'S TREE OF LIFE

Now we are well situated to think about sex in a selection context. Sex has an archnemesis: extinction. Darwin conceptually rooted this conversation in a propagational chart. Which genetic lines persist and which extinguish, and why? Darwin and some biblical authors jive in unexpected ways here. He speaks of this as ratio striving in animals:

> All that we can do is to keep steadily in mind that each organic being is striving to increase in a geometrical ratio; that each, at some period of its life, during some season of the year, during each generation, or at intervals, has to struggle for life and to suffer great destruction.[16]

---

[16]Charles Darwin, *The Origin of Species, by Means of Natural Selection*, 6th ed. (London: John Murray, 1872), 61.

Every single organic being may be said to be striving to the utmost to increase in numbers; that each lives by a struggle at some period of its life; that heavy destruction inevitably falls either on the young or old during each generation or at recurrent intervals.[17]

Darwin's tree of life unironically picks up on the biblical terminology from Eden and employs it as an analogy toward a different end. Instead of offering life to those who eat of it, Darwin's tree depicts the genealogy of various species branching out from the trunk of common ancestry. But the key to this tree diagram is that most of its branches had to go extinct. He focuses it on the persistence of only a few that can fit in the struggle for life.

The diagram intends to represent the idea that early life would produce all kinds of speciating branches. Dozens or hundreds of similar bacteria, jellyfish, or eagles would struggle to fit and fight to survive and propagate within the nutritive environments in which they found themselves. But, of course, not all of those species would survive. By dint of climates and resources, most would enter extinction. Their branch ends.

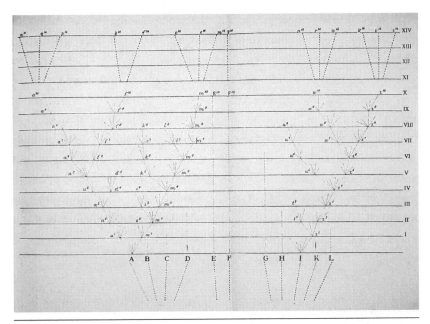

Figure 10.1. Darwin's "Tree of Life"

---

[17]Darwin, *Origin of Species*, 52.

It is unclear whether or how much Darwin intended his tree to satirize Eden's tree of life. In the next chapter, I will suppose that the conceptual girth for Darwin's tree of life finds a unique ancient companion in—you guessed it—the genealogies of the Hebrew Bible, but not in other ancient Near Eastern genealogies.

For now, it is only worth mentioning Darwin's tree of life because of how it maps out the crucial role each sexual practice plays in an unfolding genetic drama. Though Darwin toyed with several other biological analogies—coral, watershed, lungs, and blood vessels of life—the tree won out.[18] The tree chart is styled and shaped by striving to increase one's genetic advantages through propagation, but mostly ending in death and extinction of most organisms in the so-called natural history of life. Though it is called the tree of *life*, most of its foliage ultimately signals sex and then death.

CONCLUSIONS

No one knows the natural history of sex. It is not in the archaeological record, though maybe some technology will expose it eventually. Viral insertions spliced into our genes, among other complicating factors, makes an archaeology of our genetics improbable.[19] So, we are all stuck speculating about the natural history of sex.

We do know that humans are a minority in the animal kingdom as sexually dimorphic, that males and females have significantly different body mass. This difference in sex and size requires the negotiation of sexual practices. But coercive and forced sexual acts commonly occur among animals such as humans, ducks, and other species where the males grow bigger and stronger than the females. But I am not sure we can call sexual practices "good" or "bad" in the so-called natural history. Only on the most generous and ideal version of a cooperative natural selection model could sex foster exclusive coupling and avoid the problem of grounding hominin development in a long history of forced copulation.

---

[18] According to Jonathan Eisen, an evolutionary biologist at the University of California, Davis. Eisen, "Top Five Metaphors Darwin Considered and Rejected for 'The Tree of Life,'" Jonathan Eisen's Lab, June 30, 2008, https://phylogenomics.me/2008/06/30/top-five-metaphors-darwin -considered-and-rejected-for-the-tree-of-life/.

[19] S. Joshua Swamidass, *The Genealogical Adam and Eve: The Surprising Science of Universal Ancestry* (Downers Grove, IL: IVP Academic, 2019), 33-41.

Biologists today concentrate on identifying the genetic content of random and nonrandom sexual propagations.[20] Instead, I have focused on the means of propagation, as did Darwin. So too did the biblical authors.

In the coming chapters, we will see a dramatically different approach to sexual partnering across Scripture. Apart from putting biological sex (male/female) as part of God's image and at that very moment of human creation, the biblical authors show that they are neither blind nor indifferent to various sexual encounters. They control for the best version of propagation by arguing for a particularly risky form of partnering from a natural selection point of view: monogamy.

---

[20]Revisiting a prior point, most biologists' big questions about propagation today center on genetic drift, not natural selection. Genetic drift derives from random gene mutations caused by things such as solar radiation. It involves any or all the nonrandom sexual pairing practices discussed above. But the methodological question often becomes, How can we assess random from nonrandom generational genetic changes within and between species? More simply: Can we ever truly understand the connection between sexual selection and genetic expression?

CHAPTER ELEVEN

SEX AND GENERATION ACROSS GENESIS

HERE IS AN ADMITTEDLY STRANGE thought experiment among koala: What if it were the case that koala bucks (males) refused to copulate with koala does (females)? For reasons unknown, the bucks lose their interest in procreating, and hence propagating. Year after year, there are no koala joeys. Year after sexless year, the females become more aggressive with the bucks during mating season. Eventually, the strongest does work together to force copulation with the unwilling bucks. This is only a thought experiment. Yet, what kinds of judgments might we make about female koalas who make their extinction-longing bucks copulate by sheer force? What kinds of moral judgments would we make if it were women forcing men?

This is not a hypothetical case for the biblical authors. In Genesis, two women get a man drunk in order to force copulation for the stated purpose of perpetuating their genetic lineage. "Our father is old, and there is not a man on earth to come in to us after the manner of all the world. Come, let us make our father drink wine, and we will lie with him, that we may preserve offspring through our father" (Gen 19:31-32 modified). Incredibly, the perpetrators are his daughters! But notice that Lot's daughters are not worried about reproduction. Their fears stem from generational anxieties.

The ethics of sexual partnering get as complicated as we could imagine in the Hebrew Scriptures/Old Testament, continuing into the New Testament. Yet the focus of those texts remains on generation, not merely reproduction.

The biblical discussion on sexuality does not end in the Torah or Hebrew Bible. It gets gathered up in the analogies for the kingdom of God, which include buffering against scarcity and reorienting creation so that habitat and creature fit without fear of extinction.

A BIZARRE ASIDE ON MODERN SEX

For most of us today, neither reproduction nor generation leaps to our minds when we think of sex. Even our common parlance about sexuality has been recently transformed by an exceptional sense of self, centered on our desires, locked into sexual identity, and focused on sexual preference. Whether we think these are good or bad developments, the components of what we call sexuality would appear almost unrecognizable in the Hebraic thought-world (and most cultures throughout most of human history).

Right, wrong, or otherwise, we in the WEIRD world have systematically estranged ourselves from humanity's storied preoccupation with sexual generation.[1] This has left sex, in the twentieth and twenty-first centuries, predominantly entangled in matters of an individual's desire, consent, and pleasure—in that order. Surely this entanglement has some benefits and deficits. For the purposes of encountering the biblical intellectual world, our modern and niche views of sex can estrange us from how the biblical authors conceptualize the relationship of sex to generation.

Sometimes, we do stop and think in terms of generation. But most of us will not think in measured terms about sexual generation until we fear we are losing our own reproductive abilities (e.g., infertility, age) or rights (e.g., the one-child policy of China).

In the fractured world of sex we have experienced, untold numbers of women and men have been sexually abused. Assault and abuse have undoubtedly shaped our concepts of sex. Acts of forced copulation have made some survivors of such assaults themselves emotionally unable to even think about procreating and raising a family. By any angle we inspect, it seems obvious that coerced and forced copulation harms everyone involved. But by all measures, it asymmetrically harms those who have been *forced* and *coerced*.

---

[1]WEIRD is the standard acronym for Western, Educated, Industrialized, Rich, and Democratic.

This to say, generation is often not on *our* minds when we think of sex and everything that goes with it. Tangential complications of sex, such as preventing pregnancy—a reproductive concern—are about as close to generation as our thoughts might go. Of course, those who are dying, those with estates to be executed, those with land to be worked in perpetuity, the elderly who lived a life of solitude, and people struggling with infertility can be frenetically preoccupied with both reproduction and generation.

What belongs to sex? Chiseled by the habits handed to us, we think instead through a culturally stratified lens of sex cultivated in us from childhood onward. It should now be obvious that I am using the term *sex* in an oddly broad manner. By *sex*, I mean sex in the most encompassing way possible: the entire complex of habits, behaviors, appearances, rituals, and more associated with the act itself. Geoffrey Rees describes our awkward myopic sex-world this way: "Sex is not something people do, nor is it something people are. It is something that people become, a possibility of intelligible personal identity with a history."[2]

By Michel Foucault's lights, *sex* refers to "the desire to have it, to have access to it, to discover it, to liberate it, to articulate it in discourse, to formulate it in truth."[3] Sex refers to everything that contributes to the act of sex and ensues it.[4] Rees interprets Foucault to also mean that diets, plastic surgeries, pornography, anorexia, homicidally abusive relationships, and so on all indicate our willingness to contort, maim, abuse, and kill ourselves at a hare's pace in order to secure sex, love, security, identity, and intimate knowledge in all its vaporousness.[5] So too, our longings grew to secure our desired gender and sexual identity—another debated notion of recent vintage.[6] And by selling the narrative of "sexual identity" to the Western world, what Foucault calls our Faustian pact, we fortify and centralize desire and control into all things sex.

---

[2] Geoffrey Rees, *The Romance of Innocent Sexuality* (Eugene, OR: Cascade, 2011), 49.
[3] Michel Foucault, *The History of Sexuality, Vol. 1: An Introduction*, trans. Robert Hurley (New York: Pantheon Books, 1978), 156-57.
[4] Judith Butler reminds us that Foucault had generation in mind here: that wars will be waged "so that populations and their reproductions can be secured." Butler, "Sexual Inversions," in *Foucault and the Critique of Institutions*, ed. John Caputo and Mark Yount, Greater Philadelphia Philosophy Consortium (University Park: Pennsylvania State Press, 1993), 81-100, here 96.
[5] Homicide is the primary cause of death of pregnant women in the US: www.hsph.harvard.edu /news/hsph-in-the-news/homicide-leading-cause-of-death-for-pregnant-women-in-u-s/.
[6] E.g., Andrea Long Chu, "My New Vagina Won't Make Me Happy," *New York Times*, November 24, 2018, www.nytimes.com/2018/11/24/opinion/sunday/vaginoplasty-transgender-medicine.html/.

Why take this bizarre aside on modern sex and the fictions of sex that we have spun for ourselves? Whether we are Christians, atheists, or something else, we hold significantly different views of sex today from those of the ancients, medievals, and even most humans of the last few centuries. We have become estranged from sex's purpose, its social functions, its outcomes, and its psychology according to the reasoning of most humans throughout history. In their gaze, our sex obsessions and practices must seem dystopian.

We might think that this estrangement is good or needed. We might find the view of sex-as-generation too binding or fraught with potential for misogyny. But this also requires that we listen to those discourses that we value on the topic from within their own frame. We have heard a little bit on sex from Darwin and the evolutionary sciences, and now we turn to Scripture.

The difference between reproduction and generation. In light of sexual generation, the biblical authors offer the possibility that there might have been a time when forced copulation did not exist.

Genesis remains silent about animal sex inside and outside the garden, assuming that we see Genesis 1:1–2:4 and Genesis 2:5-25 as separate versions of creation. Maybe some animals were fruitful asexually, reproducing themselves. But that is not how the biblical authors seem to have thought about sexual reproduction. The second act of creation does refer to human sex, but in the context of the couple's intimacy, not generation (Gen 2:24). Though it may be naturally inferred from the union of "one flesh," the biblical author only later explicitly connects sex to reproduction, and only after the couple is exiled from the garden (e.g., Gen 4:1). Stated differently: sex has a role to play within human intimacy in league with its procreative outcome. This can even be seen in some of the biblical language for sex itself.

The verb "to know" (*yada'*) occasionally expresses "to have sex with" across the Hebrew Bible (e.g., Gen 4:1, 17, 25), which reveals the epistemological role of sex between persons: they come to know each other through sex. It also reveals that knowing entails intimacy for most biblical texts. "To have sex" is a rare use of "know" (*yada'*) in the Hebrew, but the mere presence of sex as a kind of knowing means that sex is somehow for the purpose of knowing the other person. Michael Carasik points out that *yada'* contains within it the idea of knowing by "coming closer to," which is all we need to know in order to understand how sex could be epistemological—for the sake of knowing something or someone.[7]

The knowing function of sex also helps to explain why the attempts at same-sex forced copulation in Sodom and Gomorrah are so deeply problematic (Gen 19:1-11). Despite their language—"bring them out to us, *that we may know* them"—the demand for forced copulation by all the men of the town is neither for reproduction nor for intimate knowledge of their partner (Gen 19:5). It violates the purposes of sex examined early on in Genesis: fruitful multiplication coupled with intimate knowing.

A reminder about generation versus reproduction. As I noted in the beginning of the book, I have been using the term *propagation* mainly as a generic reference, unless otherwise noted. *Propagation* has generally referred to the genealogical spreading of a line of creatures genetically related by means of sexual and asexual reproduction. At times, where noted, I have been making a technical distinction between generation and reproduction as an intention for sex.

Because it will be most relevant in the next chapter, I offer this reminder that I am using *reproduction* to refer to the act of a couple copulating and producing children within their own generation. *Reproductive* concerns focus on fertility, the process of coupling, and the proper contexts to conceive and raise children. *Generation* refers to the notion of progeny and ensuring a particular lineage.

In short, *reproduction* deals with the present production of children, whereas *generation* deals with the persistence of descendants through the process of reproduction over time. *Reproduction* worries about infertility; *generation* worries about extinction of a family name. I will use these terms in these distinct senses in this and coming chapters.

In the next chapter, we will consider where Jesus appears to erase reproduction from the age of resurrection (i.e., the eschaton). Read out of context, Jesus initially appears to say that marriage does not exist in the new heavens and new earth. Does Jesus remove reproduction, and hence generation, from the human endeavor? If so, what is the point of male and female sexes in the eschaton? Evolutionary science tells us of a possible time when sex

---

[7]Michael Carasik, *Theologies of the Mind in Biblical Israel* (Oxford: Peter Lang, 2005), 20.

was not, but does Jesus warn us of a time when sex will *no longer be*? To navigate this theological conundrum, we need a sense of generation beyond mere reproduction regarding the purposes of sex. The biblical authors are here to deliver that sense of generation.

GENERATION IN GENESIS 1–11

Genesis begins with construction of habitations and then of sexed inhabitants—male and female—whom God immediately blesses to "be fruitful and multiply" (Gen 1:22). After forming a sexed pair of humans, God also blesses them to procreate, propagate, and rule over creation: "Be fruitful and multiply and fill the earth" (Gen 1:28).

The blessing to propagate stands in contrast to the industrialized fertility cults of the ancient world. As Claus Westermann notes, "Such divination is not possible in Genesis because the Creator alone is divine and the power of fertility which is contained in the Creator's blessing is directed entirely to man."[8]

It is striking that the only notable thing about the couple being created "in the image of God," according to Genesis 1, is their maleness and femaleness—their ability to reproduce:

> So God created humankind in his image,
>> in the image of God he created them;
>> male and female he created them. (Gen 1:27 NRSVUE modified)

Even here, in Genesis 1, we find a "species" view of generation over against the later tribe- and clan-focused generation of the patriarchs.[9] It is primarily about the propagation of humanity as a species, not a particular family's lineage.

In the story of Genesis 2, Yahweh Elohim shapes the dirtling (*ha'adam*) from the dirt (*ha'adamah*) and then builds the woman (*ha'ishah*) from the man (*ha'ish*). The nature of their relationship gets brief attention, inviting the audience to imagine a couple's future independence from hypothetical parents ("[he] shall leave his father and mother") and their sexual union ("and they shall become one flesh") with no talk of kids in sight (Gen 2:24-25).

---

[8]Claus Westermann, *Creation*, trans. John J. Scullion, SJ (Minneapolis, MN: Fortress, 1974), 55.
[9]"Creation in the image of God is not concerned with an individual, but with mankind, the species, man." Westermann, *Creation*, 56.

In the tragedy of Genesis 3, the woman's reproductive organs are warped, creating strife in sexual reproduction. Again, this contortion concerns her lineage and generation, not merely her immediate children. (This specific frustration of female fertility persists into the bodies of the matriarchs of Israel and their surrogates: Sarai, Rebekah, Leah, Rachel, Bilhah, and Zilpah, though not Hagar.)

Genesis 4 spotlights three distinct acts of reproduction (e.g., "The dirtling knew Eve and she conceived and bore . . ." [Gen 4:1 my translation]). Cain's reproductive act is stated identically to the dirtling and Eve but specifically traces his generation to discrete lineages of human clans. The conceptions and births of both Cain and Abel end in murders. That first fratricidal murder is then followed by a lineage of Cain's descendants. It is also a genealogy of human industry more generally (Gen 4:20-22).

Westermann does not like the scholarly separation of Genesis 3 from Genesis 4, preferring to read them as unified. Regarding the lineage of Cain, he cites this unique use of genealogy to make a broader point about humans: "This too is a genealogy, as in chs. 5 and 10; but a genealogy of a different kind. The growth of human endeavour is presented in the form of a genealogy. It is recognized that man's work is part of the growth of the human race."[10]

Sex as reproduction punctuates Genesis 4 three times (Gen 4:1, 17, 25), but sex as generation ultimately becomes the focus when we read the sobering words at the end of this passage: "And [the] *dirtling*[11] knew his wife again, and she bore a son and called his name Seth, for she said, '*God has appointed for me another offspring instead of Abel, for Cain killed him.*' To Seth also a son was born, and he called his name Enosh" (Gen 4:25-26 modified). Notice that sex was not for reproduction in this final instance between Eve and the dirtling. Rather, their sex pointedly aimed at generation of a lineage intended through Abel, but now descending through Seth. Additionally, the Hebrew accounts portray creation in genealogical terms, commonly translated "these are the generations of . . ." (*toledot*). This generational language acts as yet another structural feature of the whole book

---

[10]Westermann, *Creation*, 27.
[11]The word traditionally translated "Adam" and which I have been translating "the dirtling" is without the definite article here. Hence, I have translated it as "dirtling" without "the" preceding it.

of Genesis, recurring eleven times within Genesis 2–37 (see Gen 2:4; 5:1; 6:9; 10:1; 11:10, 27; 25:12, 19; 36:1, 9; 37:2).

It is sexual, too, not just factive lineages. Genesis leaves little to the imagination about the varieties of sexual relations engaged in by people with and without generation on their minds. Sexual concerns for the family lineage compel what appears to be proper and improper sex acts throughout the book. So much so that a fraught map of sexual escapades may surprise the reader of Genesis who was not yet aware of this theme.

Returning to the order of events in Genesis, Genesis 5 presents a lineage as "the book of the generations of [the] *dirtling*" (Gen 5:1 modified). This genealogy ends with Noah (*noakh*) because he will bring "rest" (*nakham*) from the curse of the dirt (Gen 5:29).

In Genesis 6, readers immediately encounter a generational conflict between the multiplying "sons of Elohim" and the "daughters of men" (Gen 6:1-3 my translation). What these phrases refer to is debated. Less controversially, the story figures this generation as a cause of conflict. The official introduction to Noah begins with another instance of "these are the generations of . . ." (Gen 6:9). From the outset of the story, the author tips us off to a significant genealogical event, comparable to Seth's substitution for Abel. The headline in both instances is, "This clan, not those others."

The flood narrative unsurprisingly ends with another genealogy, but this time it maps out nations and peoples as much as an ethnographic lineage of Noah (Gen 10:1-32). We have already rehearsed the story of Babel twice, so I only want to point out one thing: Babel is part of a generational story line that largely ends in Genesis 11. These generations that descend from Nimrod and Babel become enemies later in the story of Israel.

The city of Babel (Gen 11) ends as a lesson in subverting the promise of food and stability, turning to Noah's son Shem as the line of descendants whom we will follow most closely throughout the Hebrew Bible right up through the New Testament. Genesis 1–11 is as much about origins as it is about a particularly unfolding line of sexual generation.

FERTILITY SCARCITY: GENESIS 12–50

Genesis might be the most sex-ridden literature in the Bible. There is a logic to what happens after Babel, but it will take careful attention and some getting used to. The New Testament seems to not care much about sex except for Jesus' conception and the evils of fornication. But attention to the Scriptures that Jesus knew (i.e., Hebrew Bible) shows how he relies on its generational logic to envision and participate in the kingdom of God. It is worth the effort to understand.

Genesis 12 begins a distinct story line of generation that spans the Scriptures. Notably, Abram is not promised reproduction but generation—not a child but a nation. Cycles of childbearing and threats to the lives of those children animate the drama of propagating Abram's lineage.

And while God provides grandly for the patriarchs in times of food scarcity, he coordinates a crisis of female-fertility scarcity for every generation of patriarch. That scarcity is signaled by the phrase "[God] opened her womb" (e.g., Gen 29:31). Each crisis presents an opportunity to test one's trust in God, a test that the patriarch or matriarch almost always botches, even colluding their failures at times. The story of generation in Genesis thus cycles inelegantly between crises of fertility and threats to the child's life due to enmity (Gen 16; 21), foolishness (Gen 37), or a divine test (Gen 22).

Barren means sex. Before reviewing the sexual generation of the patriarchs, we must remember that barrenness presupposes sexual activity. When the text says that a character is barren or that her womb is closed, the knowledge of that barrenness presumes an active sex life. Prior to modern medical technologies (and for most people around the world today), the primary diagnosis of infertility comes through being sexually active without prophylaxis and yet not conceiving. While we know today that men are commonly the source of infertility, Genesis explicitly lays the blame on God, who actively opens and closes wombs (see Gen 21:1; 29:31-35; 1 Sam 1:5).

Generational anxieties affect infertile couples across the biblical literature, but also widows (Gen 38:6-11) and eunuchs (e.g., Is 56:3-5). Ironically, while God makes abundant food a nomadic source of security and wealth for the patriarchs, he simultaneously afflicts their wives' and their servants' wombs with fertility scarcity, selectively opening and closing the reproductive paths of the matriarchs and their female servants.

The complications of infertility in Genesis 12–39 appear especially absurd given God's promise of a nation of descendants to Abram (whose name means "father of many") and repeated to his sons Isaac and Jacob.

Of all these instances, it is the matriarchs and other women who identify generation as their central concern. Sarai, Lot's daughters, Leah, Rachel, and Tamar are clearly motivated to carry out sex acts—sometimes undesired, forced, or prostituted sex acts—in order to reproduce for the sake of generation. Conversely, Genesis portrays the patriarchs as negligently acquiescent or inappropriately obstinate to the demands of these women.

Promises of generation. Beginning with Abram, God's promise to protect Abram and propagate him into "a great nation" is immediately followed by Abram's fearful lending out of his wife sexually to the Egyptians (see Gen 12:2, 12). Abram gains wealth from his decision to hide his identity from the Egyptians, and his wife is returned to him barren. Abraham and Isaac will both fear those among whom they sojourn, lie about their wives, and expose their wives' reproductive organs to sex acts with foreign leaders. Each time, Abraham and Isaac gain wealth and secure food from the transaction, but never a child of the divine treaty (see Gen 20:1-18; 26:6-16).[12]

Hang with me here. The only way to see the conceptual constellation of sexual generation emerge from the starry skies of Genesis is to get into the nitty-gritty of story flow.

God reappears to Abram and reaffirms his promise of children as numerous as the stars, which Abram trusts (Gen 15:5-6). God also promises all the land from Egypt to Mesopotamia, about which Abram is skeptical, asking, "How am I to know that I shall possess [the land]?" (Gen 15:8).

Immediately after a story in which Abram trusts the promise for children, Sarai hatches a plan to reproduce children through her servant Hagar. Ancient and modern commentators struggle with whether this was an appropriate arrangement. I will only shine a light on the language of this story in Genesis 16. First, the exact wording of the transaction is reminiscent of the garden error. Eve "took" the fruit and "gave" it to "her husband," which ends in knowing good and evil (Gen 3:6). The problem identified by God in Eden was the man's error: "because you have listened to the voice of your wife," reminding the dirtling that he and she were both there listening to the serpent (Gen 3:17; see Gen 3:6).

---

[12]At some point, with Abraham at least, the second instance of handing his wife over to a foreign ruler appears to border on prostitution, especially if he has any notion of gain in mind (and it would be historically justifiable to suspect material wealth as an outcome at that point).

Table 11.1 . "Listen to the voice" in Genesis

	Gen 3:6/16:3	Gen 3:6/16:3	Gen 3:6/16:3	Gen 3:17/16:2
Eve	"took" (*laqakh*)	"gave" (*natan*)	"to her husband"	"for you have listened to the voice of your wife" (*shama' leqol*)
Sarai	"took" (*laqakh*)	"gave" (*natan*)	"to Abram her husband"	"Abram listened to the voice of Sarai" (*shama' leqol*)

In parallel to Eden, Sarai "takes" her servant Hagar and "gives" Hagar to her husband so that he can know her (Gen 16:3). The narrator includes that phrase from Eden, "Abram listened to the voice of Sarai" (Gen 16:2). This scene so closely resembles Genesis 3:6 that biblical scholars have seen Sarah's actions with Hagar as a reenactment of Genesis 3.[13] The unmissable and ironic through line is spectacular in its tragedy: Sarai's barrenness, which descends from Eve's curse, is also the event that she reenacts with Hagar as the fruit taken and given to her husband.

When Sarah then plans to deal with Hagar and child, Abraham again listens to the voice of Sarah. This time it is at God's direction, and he maroons the child of that union with Hagar to die in the wilderness. Though the child's life is threatened, God reassures Abraham that Ishmael will be saved and will become a nation (Gen 21:12-13). Generation through Hagar is secured, even if it turns out to be perpendicular to the central story line of Shem's line through Abraham. Hagar's branch then extends offstage, as it were, through Ishmael—a story line not pursued by the biblical authors.

What about Sarah's line? God promises Sarah a lineage through Isaac, the laughable child to come. But before Isaac arrives, we encounter another set of women worried about progeny and extinction.

Sex to avoid extinction. Lot and his daughters escape the city where Lot had just offered them up to be gang-raped by all the men of Sodom (Gen 19:8). Now hiding in a cave, his daughters realize that Lot's wife and their own future husbands are all dead. Their own Faustian pact seems to indicate that they thought God might have destroyed all humanity except them: "Our father is old, and there is not a man on earth to come in to us after the manner of all the earth" (Gen 19:31). Hence, they might have

---

[13]Gordon Wenham, *Genesis 1–15*, Word Biblical Commentary 1 (Nashville: Thomas Nelson, 1987), 8; Werner Berg, "Der Sündenfall Abrahams und Saras nach Gen 16:1-6," *Biblische Notizen* 19 (1982): 7-14.

perceived this as another postflood situation from within the logic of the narrative. To bolster this explanation, the only other parallel remotely similar to this also involves two women agreeing, this time to feed cannibalistically on their infant progeny (2 Kings 6:24-31). This suggests that Lot's daughters might have legitimately thought this was an end-times event, which might help to explain what happens next. Like the two cannibalistic mothers, these extreme behaviors reveal the total desperation of scarcity mindset.

The two daughters agree to get their father drunk and coerce him into sex with them. They state their reasoning clearly: "There is not a man on earth . . . we will lie with [our father], *that we may preserve offspring from our father*" (Gen 19:31-32).

Not only do they carry out the deed, but the narrator uses this story genealogically to show that generational concerns are answered in the nations of Moab and Ammon (Gen 19:36-38). In Lot's case, it is probably safe to remain agnostic on whether this was an instance of forced copulation by his own daughters. Their actions were more coercive than forced, but explicitly for the sake of generation.

Extinction by child sacrifice. That is what was happening in Lot's family. What about Abraham and Sarah? Sarah finally conceives and delivers the laughed-at son of the covenant: Isaac. Again, Abraham's son is born and then imperiled. This time, it is God who threatens Isaac's life in a test of Abraham. In asking for an offering, the story of the attempted sacrifice of Isaac exacerbates the absurdity of God's test by hyperbolically repeating familial language. Almost every mention of Isaac or Abraham in this short story includes a "relational epithet," repeated a dozen times over fourteen verses (as if the reader needed to be reminded of how these two people were related to each other, or that their relationship was of particular importance).[14] For example:

- his father Abraham (Gen 22:7)

- his son Isaac (Gen 22:3)

- "My father!" . . . "Here I am, my son" (Gen 22:7)

- "your son, your only son" (Gen 22:2)

---

[14]Robert Alter would also want us to notice the narrator's immoderate repetition of "relational epithet" in this story. Alter, *The Art of Biblical Narrative*, rev. and expanded ed. (New York: Basic Books, 2011), 224. See Gen 22:1-3, 6-10, 12-13, 16.

Most presume that these exaggerated restatements of their father-son relationship intend to highlight that it is the lineage, not just Isaac, being sacrificed.

The pattern of barrenness, pregnancy by divine womb-opening, and then threat to the child's life culminates on Mount Moriah, where Abraham steadfastly assures Isaac, "God will provide" (Gen 22:8). The narrator spotlights God's provision with the naming of the place "Yahweh provides" (or "sees to it"). A later editorial voice piles on to explain that this scene is why Abraham's multitudinous progeny say "to this day, 'On the mountain of Yahweh, it shall be provided'" (Gen 22:14 my translation).

Secret prostitution for generation. The last instance of patriarchal generation comes from Tamar, the widowed wife of Judah's son Er. God widows Tamar by killing Er for his wickedness. Widowhood made life precarious for women, but later teaching in the Torah allows widows of childbearing age to propagate their lineage through the next of kin (see Deut 25:5-10; Ruth 1:11-13). Er has two brothers who could father Tamar's children and propagate their brother's name. This appears to be a sort of socio-coercive copulation meant to be enforced by the father of the clan.

Judah cooperates for the sake of his own lineage through Tamar. He identifies lineage as the reason and the goal. He tells his son Onan, "Go in to your brother's wife and perform the duty . . . and raise up offspring for your brother" (Gen 38:8). However, Onan intentionally ejaculates on the dirt "so as not to give offspring to his brother" (Gen 38:9). So, God kills Onan for his wickedness (Gen 28:10). Then Judah instructs Tamar to wait for another brother, Shelah, to come of age. However, the narrator conceals Judah's mixed motivations for withholding Shelah from his daughter-in-law. He never arranges conjugal visits for Tamar with Shelah.

Notice that God acts on behalf of Tamar, killing Onan for cutting off *her* progeny. Tamar then disguises herself as a prostitute on the presumption that Judah likes to visit prostitutes. She deceptively gets Judah to copulate with her, producing a child unbeknownst to him (Gen 28:12-26). Notice also Judah's reaction when he discovers that Tamar has become pregnant by tricking him: "She is more righteous than I" (Gen 38:26). This reaction perplexes many readers, but only because it sneaks their modern assumptions about sexuality into their thinking. The fundamental sexual desire to propagate one's lineage can seem puzzling. This cultural blind spot hinders us from seeing the security and labor that children could provide to widowed mothers living in subsistence farmsteading in a dangerous land.

In contrast with these instances, we can think of Genesis's depiction of coerced copulation by Potiphar's wife with Joseph, the only woman in Genesis who makes sexual advances that a man refuses (Gen 39:10). Even here, the narrator carefully flags the reason for her coercive copulation attempts: "Joseph was handsome in form and appearance" (Gen 39:6). She seemingly had neither reproduction nor generation in mind.

The narrator of Genesis treats women's sexual desires with maximum generosity, at least in the form of moral silence in the sobering depictions of these stories. As long as women aim at fruitful generation of the covenantal children, then the narrator extends patience. However, Sodomite forced copulation and the Egyptian woman's lust appear to be problematic because they do not aim at reproduction or generation of Abram's family.

A baby-making contest extraordinaire. Finally, a discussion of sexual generation in Genesis would be incomplete without briefly addressing the baby-making competition between Jacob's sister-wives. By the end of the contest, twelve boys and one girl emerge from the wombs of four women to establish the twelve tribes of Israel. However, by the end of the exile, only two tribes will survive: Judah and Levi.[15] It is a story of diverse copulation, sometimes coerced through financial means (e.g., when Rachel prostitutes Jacob to Leah to produce Issachar, Gen 30:15-16).

These two sisters, Leah and Rachel, compel Jacob to have sex with their servants and themselves seemingly for the sake of reproduction. The name of each son refers to the victory within the child-making competition. Most of those sons' names perversely reflect the focused determination of the mothers to reproduce with Jacob (e.g., Issachar: "whore's wages"; Naphtali: "wrestled [my sister] and won"; etc.). Though these women feverishly attempt to reproduce with Jacob, the narrator carefully reveals that God controls their generation, opening and closing their wombs. Their intermittent barrenness presumes a constant stream of sexual activity between Jacob, his wives, and their servants.

---

[15]The tribes of Benjamin and Simeon appear to fold into Judah and make some minor cameo mentions in Second Temple Judaism. Judah and Levi are the major characters whose story lines we follow.

All in all, their desires to reproduce with Jacob sometimes hints at a desire for generation. If they merely wanted to procreate, Jacob's semen is one of many options. But the two sisters want to reproduce *with Jacob* and fight sibling relationships in order to propagate his genes.

WHAT IS GOING ON HERE IN GENESIS?

In all of these scenes, generation appears again and again as the concern for the women, but strangely not for the men, the ones who received the promise of lineage from God in treaty form. There is a generalized concern for males regarding generation and extinction (e.g., one's name being cut off from the face of the earth), but nothing akin to the women's roles in Genesis. And when reproductive goals are absent, the sexual context is depicted negatively (i.e., Sodom, Shechem's prince, and Potiphar's wife). But even when generation guides the sexual behavior, it can be depicted as foolish at worst or righteous by a whisker at best.

Lot's daughters sought to take matters into their own hands, and Tamar did as much in desperation for generational justice from her father-in-law. Judah's reaction reveals the best-case scenario for such attempts: "She is more righteous than I" (Gen 38:26). The reader notes Judah's relative comparison: Tamar is more righteous than the man who would not honor her requests for sexual generation but likes to visit prostitutes when going out to the fields—a habit that seemingly allowed her to entrap him (Gen 32:15)!

CONCLUSIONS FROM GENESIS

The divine hand in generation and coupling, not mere reproduction, appears as a drumbeat across Genesis. No doubt, God propagates his people, with whom he has made a treaty. However, in a yet-burgeoning view of sex in the Bible, no positive assessment can be squeezed from the various and concocted human plans to propagate. God generates entire people groups from sexual assaults. The people of Moab and Amon generate from two daughters sexually assaulting their father (Gen 19:30-38). The tribes of Issachar and Zebulun (and Dinah) generate from Rachel prostituting Jacob out to Leah for the price of mandrakes (Gen 30:14-18). These all appear to be guided by God's opening of wombs over time, beyond the first act of reproduction, and through discrete lineages.

In the final assessment, the lineage of just two of Leah's sons survives the tests of the Mosaic covenant. Only Judah and Levi, sons from Jacob's unwanted first wife—a woman with whom he never intended to copulate—provide the genetic basis for John the Baptist, Jesus, and all of his apostles (as far as we can tell).[16] Along the way, unexpected non-Hebrew genes are folded in through faithful female "strangers" who immigrate from the nations, such as Rahab the Jerichoan and Ruth the Moabite.

Many have noted that Jesus' genetics include those of three other women, all of whom participate willingly or not in coercive sex: Tamar, Bathsheba, and Ruth (so some think). Regardless of whether this assessment sticks, it raises the basic point in the Hebrew Bible: generation is guided by God and takes unexpected paths, which seems to mitigate any mechanical view that would praise coercive attempts at generation as treaty fulfillment. As we will later discover, sexual generation across Scripture always includes the planned incarnation of Jesus. Hence, sexual generation beyond reproduction remains a main feature of biblical thinking into the new covenant.

---

[16]It's complicated. As Jason Staples has recently argued, first-century Judaism sometimes depicted the missing ten tribes as troops in waiting from Babylon. Neither are all Jews in Judea from only the tribes of Judah and Levi (e.g., Paul is from Benjamin [Rom 11:1; Phil 3:5], Anna is from Asher [Lk 2:36], and other cases such as the Samaritan woman). Staples examines texts that indicate the complexities of the concepts of "Jews," "Judaism," "Judea," "Israelite," and other mitigating factors in Second Temple Jewish identity. Staples, *The Idea of Israel in Second Temple Judaism: A New Theory of People, Exile, and Israelite Identity* (New York: Cambridge University Press, 2021).

CHAPTER TWELVE

GENERATION ACROSS THE HEBREW BIBLE AND NEW TESTAMENT

SEX'S STORY LINE PERSISTS and expands conceptually across the Scriptures. I promise you that this will not get boring, but it will demand our full attention. So, let's keep going.

We continue to trace the burgeoning conceptual development of sex, reproduction, and generation beyond Genesis. What we discover is an impractical moral grounding of sex, but only by considering how various sexual practices and prescriptions work together in the Torah and beyond. In the New Testament, sexual generation continues. Humanity's blessing to generate—be fruitful and multiply—becomes the metaphor for propagating disciples and the empire of God.

Leaving Genesis, female fertility no longer poses a problem. The Hebrews are now fruitful and multiplying (Gen 47:27; Ex 1:7). That harrowing discourse on sex begun in Genesis continues into the exodus, but many of the sexual questions raised in Genesis will be clarified later in the Torah's instructions.

Something beyond the mere genetic persistence of Hebrew lineage emerges as central to generation. The moral life of Israel animates the biblical theme of generation. Those who live out the justice of Torah in fear of Yahweh are the ones whom God will generate, even when their personal failings are replete (e.g., Abraham and David).

GENERATION BECOMES NATIONAL

Israel is no longer consumed with generation of the twelve tribes. Now what? By the opening of Exodus, the patriarchs' fertility crises have ended. Continuing the theme of scarcity and plenty, the sexual metaphor of being "fruitful" (*parah*) functions as the first conflict in Exodus, associated with both Eden and the promises of a nation to Abraham. Exodus redeploys the theme of fruitfulness and flourishing: "But the Israelites were fruitful and prolific; they multiplied and grew exceedingly strong, so that the land was filled with them" (Ex 1:7 NRSVUE).

The Hebrews' fruitfulness itself generates a conflict in the story. The first Pharaoh of Exodus sees Hebrew multiplication not as a potential good for Egypt but as a growing threat to Egypt's security (Ex 1:10). Israel's fruitfulness creates a wrinkle that Pharaoh's infanticide means to iron out.

But on behalf of the Hebrews, female civil disobedience surrounds the Pharaoh who works the Hebrews oppressively and orders the murder of their boys.[1] First, the midwives Shiphrah and Puah civilly disobey Pharaoh's orders to murder Hebrew boys at the birthing stool (Ex 1:15-21). Second, the Pharaoh then commands *all* Egyptians to murder Hebrew boys, the results of which go unstated by the narrator (Ex 1:22). Third, Moses' mother civilly disobeys and hides the child to be satirically thrown into the Nile in a basket—a story that follows an ancient pattern of infant exposure that ends in the thunderous return of the threatened child. Moses' sister also plays a complicit role in their disobedience.

Finally, Pharaoh's daughter herself disobeys, completing the circuit of the renowned infant-exposure motif. Like Luke Skywalker, Mesopotamia's Sargon I, the Egyptian god Horus, or J. K. Rowling's Harry Potter, the hunted child of promise eventually returns to dominate the powers that sought to snuff the child out in their infancy.[2] By entering that well-worn story line, Pharaoh's daughter ironically becomes the one who raises Egypt's future

---

[1] Yoram Hazony first pointed out this pattern of civil disobedience to me in a conversation.

[2] Horus was abandoned among the reeds of the Nile to be raised in obscurity and returned as a triumphant adult. Similarly, Sargon was abandoned in the Euphrates and returned. Donald B. Redford, "The Literary Motif of the Exposed Child (Cf. Ex. ii 1-10)," *Numen* 14, no. 3 (November 1967): 209-28. In some Mesopotamian traditions, infant exposure appears to be a legal step in the hope of adoption. Legal texts sometimes cite that the infant was left out "to the dogs" as the legal basis for the adoption. Shawn W. Flynn, *Children in Ancient Israel: The Hebrew Bible and Mesopotamia in Comparative Perspective* (New York: Oxford University Press, 2018), 87.

terrorizer, the very sort of Hebrew whom this unnamed Pharaoh sought to kill. (This motif repeats in the Gospels, where a "Jewish" king Herod now seeks to kill a predicted usurper in infancy. But in the Gospels, Egypt becomes the refuge from the murderous king.)

The biblical authors fuse the reproductive abundance in Exodus with political uncertainty and murderous population control. Between Egypt and Canaan, generation, politics, and inheritance of land interweave up until the exile. Discussions of land inheritance come to the fore as Israel prepares to settle Canaan, but all inheritance talk includes the matter of generation. In the wilderness wandering, the daughters of Zelophehad lose the males of their family. They plead with Moses and later Joshua in order to inherit their portion of land with their brothers, presumably so they can generate the family name (see Num 27:1-11; Josh 17:3-4). Notice that land, which signals the ability to produce food and survive, is tied directly to generation. The daughters look beyond scarcity and reproduction to the horizon of sufficiency and descendants.

A later property dispute in the divided kingdom also turns out to be a matter of land for the sake of one's progeny. Naboth's answer to King Ahab's attempt to buy his land reveals the primacy of his inheritance and its generational goals in his thinking (1 Kings 21). Not only does the property belong to Naboth, but also his family belongs to a particular plot of land. His generational concerns justify his refusal to sell to the king but also set in motion an assassination plot to steal his land. In the Hebrew logic of generation, it is not just his land that is stolen, but more importantly his inheritance for his children's children. This act is deemed so heinous that God plots Ahab and Jezebel's assassination in response. This inheritance theft also reverberates directly into Micah's diatribe against Israel as a whole:

And Ahab went into his house vexed and sullen because of what Naboth the Jezreelite had said to him, for he had said, "*I will not give you the inheritance of my fathers.*" And *[Ahab] lay down on his bed* and turned away his face and would eat no food. (1 Kings 21:4)

> Woe to those who devise wickedness
>   and *work evil on their beds*!
> When the morning dawns, they perform it,
>   because it is in the power of their hand.

> They *covet fields and seize them,*
> and houses, and take them away;
> they *oppress a man and his house,*
> *a man and his inheritance.* (Mic 2:1-2)

The connection between a group's morality and its generation stems from the Torah. Deuteronomy's promises about the future mix field and female fertility with the Edenic language of fruitfulness. If Israel listens to the voice of Yahweh and enacts the justice of Torah across Canaan,

> all these blessings shall come upon you and overtake you. . . . Blessed shall you be in the city, and blessed shall you be *in the field.* Blessed shall be the *fruit of your womb* and the *fruit of your ground* and the *fruit of your cattle,* the *increase of your herds* and the young of your flock. (Deut 28:2-6)

By this depiction of how well things could go and the ensuing horror show of how grotesquely God could punish them, Deuteronomy 28 taps deeply into the fears of agrarian scarcity and generational anxieties all at once. Starvation both kills and cuts off one's genealogy in the land.

All of Israel's fruitfulness and multiplication—her prosperity—can be reversed. What once flourished could be metaphysically reoriented toward frustrating every Hebrew endeavor:

> But if you will not obey the voice of the Yahweh your God . . . then all these curses shall come upon you and overtake you. Cursed shall you be in the city, and cursed shall you be *in the field.* Cursed shall be your basket and your kneading bowl. Cursed shall be the *fruit of your womb* and the *fruit of your ground,* the *increase of your herds* and the young of your flock. (Deut 28:15-19 modified)

Enumerated and horrifically depicted, these curses reveal some of the hardest prose of Christian Scripture to read. As previously mentioned, God promises to starve them to the point of secret and selfish cannibalism of their own children, which eventually comes true in the northern nation of Israel in 2 Kings 6:24-31: "And you shall eat the fruit of your womb, the flesh of your sons and daughters" (Deut 28:53). Their wives will be raped in front of them by foreign invaders (Deut 28:30). Their children will be given to another people (Deut 28:32). Their animals will be taken, and they will ultimately be exiled from the land (Deut 28:36). In other words, they will be terrorized and then eliminated if they do not embody the justice of Torah in the land for the sake of the vulnerable and great alike (Lev 19:15).

As it was for the Canaanites before them, this exile will be the inevitable outcome of Israel's incessant nurturing of injustice across their land. Their ability to generate the nation is bound to their willingness to care for their vulnerable, flocks, fields, and foreigners. Israel's slump toward endemic injustice causes the extinction of ten branches, plus or minus, on the northern Israelite tree of life. Their survival depends on a relatively low bar of moral behavior. Eventually, Judahites will succumb to oppressing their poor, and their kings will sacrifice Hebrew children to other gods, forcing the same divine slaughter and exile suffered by the Canaanites and Israel to the north.

We can now see the thread sewn throughout the history of the Hebrews. Generation and being fitted to the land depend on being a certain kind of people—wise and discerning in their proliferation of justice and mercy.

NORMATIVE SEXUAL RELATIONSHIPS

What is a proper sexual relationship, and what do sex and generation have to do with Israel's ruinous history? By the time readers arrive at Leviticus 18 and Leviticus 20, they have seen a confusing collection of sexual encounters, including

- forced/coerced surrogate sex (Gen 16:1-6)

- attempted public male gang rape (Gen 19:1-29)

- getting one's father drunk for generational sex (Gen 19:30-38)

- prostitution of one's wife in exchange for safety (Gen 20)

- Israelite orgies to other gods (Ex 32:6)[3]

And these are merely the highlights! What we lack from the biblical authors in this morass of sexual abuse and calculations is any kind of clear guidance on what makes a proper sexual pairing. That would not be problematic if the biblical texts were not shot through with specific guidance on so many other matters of life, including inappropriate sexual partnering. But no

---

[3]The seemingly innocuous phrase "rose up to play" indicates sexual immorality, or orgies, every-where else it occurs in the Hebrew Bible, which explains why Paul and other Second Temple Jews interpret it this way (1 Cor 10:7).

constructive account of sex is ever given, at least not the way we typically think of such commands.

The most basic questions go unanswered if we naively search for direct commands: Whom should an Israelite have sex with? Under what conditions? And to what end? Instead, we find detailed lists in Leviticus of inappropriate sexual relationships with relatives and nonrelatives. These lists aim to construct paradigms of reproduction that will support the generation of the tribe.

In the first section of Leviticus 18 (Lev 18:6-18), sexual contact with "close relatives" is forbidden.[4] Of course, in a foundational story that contains a man and his half sister as the parents of all Israel (i.e., Abram and Sarai), the question must be asked: What makes someone a "close relative"?

The answer comes in the form of a long list of kinds of close relatives. That Leviticus—in a list of almost twenty types of close relatives—does not forbid sex with one's grandparents, for example, seems to indicate that this list intends to build paradigms rather than to communicate a legislative approach to sexual behavior.[5] In other words, such "laws" are not statutory law as we think of them commonly today: rules that name every offense and are either upheld or broken. Rather, they create concepts and case studies to instruct Israel in their discernment. There is a way that Israel should exist as a nation that cannot be captured or exhaustively detailed in statutes as we think of them today. The legal reasoning of the Old and New Testaments is not interested in such a statutory project of enumerating laws and conditions to be broken or maintained.

Hence, the lists are exhaustingly long, but not exhaustive, and even point to other stories in the literature. For instance, it is difficult to read the final example of a "close relative" and not think of exactly to whom it is referring: "And you shall not take a woman as a rival to her sister, uncovering her nakedness while her sister is still alive" (Lev 18:18). This forbidden relationship

---

[4]The Hebrew phrase *'ish 'ish 'el kol she'er* (lit.: "man man unto all family") is translated as "close relatives" here (Lev 18:6). The list makes clear that this phrase means something akin to "close relatives."

[5]Joshua Berman, among others, has argued that statutory concepts are anachronistic and blind the reader to how these "laws" function epistemologically in the texts and Israelite society. Laws were not held as statutes in the way we think of them in British or American tort law systems. Rather, biblical laws "were prototypical compendia of legal and ethical norms." Berman, *Inconsistency in the Torah: Ancient Literary Convention and the Limits of Source Criticism* (New York: Oxford University Press, 2017), 116.

might serve as a moral critique of Jacob's marriage to Rachel, almost none of whose descendants survive to the end of the Hebrew Bible. This rhetorical move of legally alluding to a narrative or collection of narratives—more common than we realize in Hebrew law—also indicates that this law is no mere list of rules to avoid breaking.[6]

In the second section of Leviticus 18 (Lev 18:19-23), sex with a menstruating woman, nonspouses, animals, and same-sex humans is likewise forbidden. Oddly, a prohibition against child sacrifice pops up in the middle of these sexual prohibitions, without context or interpretation.

All of this raises a crucial question: Why not participate in these prohibited sexual liaisons? Or, what ties these sexual prohibitions together? One answer is tempting: these sex acts are (mostly) not procreative, and ancient folks knew that. However, adultery can be reproductive, so reproduction is not the only matter on the table. Maybe we could say that these sexual behaviors are either nonreproductive *or* create the possibility of unwanted children (i.e., outside one's marriage), which might be yet another instance of a law alluding to Abram and Sarai with regard to Ishmael. They are either not concerned with reproduction (e.g., bestial sex acts), or, if procreative, not concerned with the generation of a lineage (e.g., adultery).

Within the logic of the legal material, other concerns surely fund the prohibitions against adultery. But in this instance, a particular complication of adultery appears to be in view. Across the ancient Near East and still today, there is no more vulnerable population than undesired children.[7] As with Ishmael, unwanted children might eventually end up in a fatal circumstance when the parents do not have generation in mind.

Though this seems tangential to the point, it connects matters of generation and scarcity by explaining why child sacrifice appears lodged in among sexual

---

[6]Legal scholar Jonathan Burnside sees the narrative shape of legal instruction in the Torah's commands for asylums (i.e., cities of refuge) in the land of Israel. His claim depends not only on the connections between particular laws and the story of Israel (e.g., the exodus and Lev 19:34) but also on the literary and rhetorical structure of the laws in their canonical form. Burnside, "Exodus and Asylum: Uncovering the Relationship Between Biblical Law and Narrative," *Journal for the Study of the Old Testament* 34, no. 3 (2010): 243-66. See also Burnside, *God, Justice, and Society: Aspects of Law and Legality in the Bible* (New York: Oxford University Press, 2010).

[7]Flynn effectively demonstrates that child sacrifice in the Levant highlights both the high value of children and the absurdity of the sacrifice. It seems to establish a system that would demand a steady supply of unwanted children (*Children in Ancient Israel*, 111-56).

prohibitions. Unwanted children were conceivably the easiest children to sacrifice, sociologically speaking. And considering the realities of agrarian subsistence life in Iron Age Israel, it is difficult to conceive of child sacrifice as a coherent act apart from the availability of unwanted children.

If these sexual prohibitions in Leviticus are intended to form paradigms of forbidden sexual contact without having to state every kind of sex that fits the paradigm (e.g., sex with one's grandmother), then from where do we get the "traditional view of marriage" in Scripture? How did we land on marriage between two opposite-sexed persons who are not close relatives, especially since the idea of a man and wife monogamously propagating is neither commanded nor described in the legal material (or almost anywhere else in the Torah, Prophets, or Gospels)?

Recalling the intellectual world of the biblical texts, we can see the authors hyperlinking later sexual texts with earlier ones. These sexual contact laws seem to presume that Genesis 2 provides the arch-paradigm for sexual contact: one man and woman commissioned to propagate. That is the *genus* of normative sexual contact (i.e., what it is), and it appears nodal (basically true) across the instruction of the Torah. The *differentia* of sexual relations (i.e., what it is not) appears in long, paradigm-building lists of prohibition, including later prophetic chastisements. As I have said, Genesis 2 takes a primary position in Jesus' thinking about sex.

Sex, then, derives from the garden at Eden, and aberrant sex gets illustrated in the patriarchs, and then instructed against and later chastised by legal frameworks and the prophets. Notably, these forms of aberrant sexuality, even between a married couple, appear to be singled out not only as failures to reproduce but also as failures to produce children who would propagate the family genetics (as opposed to producing vulnerable children outside marriage, who would be more susceptible to abandonment or death).

Sex fitting to male and female bodies has some regard for reproduction that will propagate those who fear the God of Israel. Nevertheless, appropriate sexual reproduction does not seek merely to propagate Hebrew genetics as opposed to a foreigner's. We see unflinchingly positive portrayals of intermarriage with immigrants who attach themselves to Yahweh worship (e.g., Zipporah, Rahab, or Ruth). The negative consequences for generation with the nations within and surrounding Canaan only stem from idolatry, intermarrying with those who have yoked themselves to Baal (e.g., Num 25:1-9; 1 Kings 11:1-8; Neh 13:23-27).

Thus, we now see that something apart from the mere persistence of a Hebrew genetic line has emerged as an animating force behind the biblical theme of generation: those who will live out the justice of Torah in fear of Yahweh are the ones whom God will cause to generate.

BIBLICAL EXTINCTION

Before dealing with some of the most difficult passages in the New Testament, let's briefly remind ourselves of the centrality of progeny in the Psalms. Though hardly any of us think about human extinction apart from our own personal death, the topic grows more palpable as we get older and family members pass. Jon Levenson reminds us, "In a culture in which identity is so deeply embedded in family structures, [eternal] life is . . . largely characterized by the emergence of new generations who stand in continuity and deference to the old."[8] Such reflections materialize in the stated fears of Hebrews in their psalms. But we find the psalmists crying out not only against death but also extinction in the Psalms:

> The face of Yahweh is against evildoers,
> > *to cut off the remembrance of them* from the earth.
> > (Ps 34:16 NRSVUE modified)
> May his *posterity be cut off*,
> > may *his name be blotted* out in the second generation. . . .
> Let them be before the Yahweh continually,
> > and may *his memory be cut off* from the earth. (Ps 109:13, 15 NRSVUE modified)
> Like those forsaken among the dead,
> > like the slain that lie in the grave,
> like those *whom you remember no more,*
> > for *they are cut off from your hand.* (Ps 88:5 NRSVUE)
> For those blessed by Yahweh *shall inherit the land,*
> > but *those cursed by him shall be cut off.* (Ps 37:22 NRSVUE modified)

This final excerpt highlights the link between progeny and land inheritance, which secures one's remembrance by Yahweh. Indeed, the history

---

[8]Jon D. Levenson, *Resurrection and the Restoration of Israel: The Ultimate Victory of the God of Life* (New Haven, CT: Yale University Press, 2006), 170.

of the Hebrew Bible ends with ten of twelve tribes "cut off," exiled and scattered, or indiscernible from the nations and therefore lost to the generation project begun in Genesis 1.

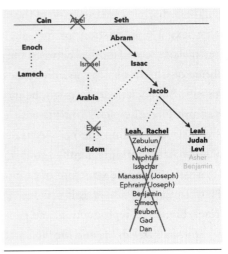

In other words, Genesis through Kings depicts the flourishing generation of a people group branching out by tribes and clans. However, by the end of their national story, that diverse tree of Israel has been pruned down (mostly) to two

Figure 12.1. The Tree of Genesis Covenants

surviving tribes deemed suitably fit to the land of Israel. (One could reasonably include Benjamin and Simeon in with Judah.)

GENERATION IN THE NEW TESTAMENT

To cut a long discussion short, the generational impulse in the Hebrew Bible generally flows from a biological emphasis into disciple-making rhetoric in the New Testament. As with many topics, silence on the matter of sexual generation in Israel appears to presuppose its continuation. In the New Testament's emphasis on what is new, making disciples overshadows making babies, because the latter persists as expected.

The generational emphasis of the Hebrew Bible occurs in precisely one portion of the New Testament narratives: the lineages of John the Baptist and Jesus. These two men and their multiplying disciples emerge as genealogically necessary from the entire lineage of the Old Testament.

The need for Israel, but more specifically for Judah and Levi, to generate their lineage appears to have been fulfilled in the messianic prophet Jesus. In turn, Jesus' focus largely aims at propagating disciples as anthropocentric ("fishers of men"), teleological ("the end of the age"), and geographical ("the end of the earth"; see Mt 4:19; 28:20; Acts 1:8). Maintaining the fruitfulness metaphor, disciples are sowers and harvesters, servers of miraculously abundant food, and containers of living water.

These disciples are the bride of the Christ, an analogy that also assumes that sexual reproduction for the sake of generation is the natural consequence of the relationship between the divine Messiah and his people. The "bride of Christ" analogy usually causes us moderns to think of the bride's beauty. But ancient brides might be more noted by their fertility: the bride can produce disciples.[9] This fits with Isaiah's renowned visions of the nations streaming into the new Jerusalem, where even the sexually bereft eunuch has a hope greater than sons or daughters (see Is 56:2-8; 66:19-21).

Again, the whole structure of disciple generation neither is new in the New Testament, nor does it presume the cessation of opposite-sexed marriage, biological reproduction, and genealogical generation.

Darwin's tree of life and the biblical end of sex. Darwin's tree of life bears a striking conceptual resemblance to the metanarrative of Israel: a proliferating variation on one line ultimately trimmed to those most suitably adapted to the environment.[10]

Additionally, Israel's twelve-tribe tree of life exhibits a unique use of genealogy in the ancient Near East. This is not how Egyptians or Mesopotamians (or Greeks!) used genealogies.[11] For instance, Egyptians tended to use genealogies to establish the pedigree of priests and prophets or to establish a particular family's right to a political office.[12] (That is precisely what some scholars think the so-called Priestly literature of the Hebrew Bible is designed to do.) One reason for the biblical authors' preoccupation with generation might stem from the very real possibility of their contemporaries within Israel immorally turning into the ill-fitting branch that withers and fades from Israel's tree of history.

---

[9]Thanks to Jon Boyd for helping me to connect these dots.

[10]It is interesting and beyond my ability to explain how Darwin's tree of life and Israel's tribal extinctions came to such conceptual parity. I am not trying to make that case. This idiosyncratic style of representing propagation and survival is notable because, yet again, Darwin's concepts appear to be Hebraic in character, even if he did not derive it so.

[11]Greek genealogies of the pantheon branched out, but no survival and pruning of the gods drove their national narrative. Given the comparative literature of the ancient Near East, Richard S. Hess concludes that there simply is no parallel to Israel's use of genealogy. Hess, "The Genealogies of Genesis 1–11 and Comparative Literature," *Biblica* 70, no. 2 (1989): 241-54.

[12]See Robert K. Ritner, "Denderite Temple Hierarchy and the Family of Theban High Priest Nebwenef: Block Statue OIM 10729," in *For His Ka: Essays Offered in Memory of Klaus Baer*, Studies in Ancient Oriental Civilization 55 (Chicago: University of Chicago Press, 1994), 205-26; Karl Jansen-Winkeln, "The Relevance of Genealogical Information for Egyptian Chronology," *Ägypten und Levante* 16 (2006): 257-73.

Does Jesus not also teach a kind of Darwinist tree of life by erasing Israel's genealogy from the kingdom of God? No! Instead of aiming at strategic extinctions, the eschaton might be pictured as perpetual branching without extinguished branches. Paul uses the grafting into a tree as a metaphor for his tree of life (Rom 11:17-24).

Does sexual generation persist beyond the resurrection? Possibly. If sex ends and generation ceases to be the goal, then the branches of the tree just keep going—with no new branches and no pruning of ill-fitted branches. Or maybe the metaphor could be read this way: only the properly fitted branches will be selected for the age of resurrection, and the ill-fitted will be the pruned branches. Either way, what is the final design for sex in the eschaton—for both the biological sexes and the sex they have?

Does Jesus cancel sex in the end? In a confusing snap back to a group of Sadducees, Jesus appears to nullify marriage (and therefore generation) in the age of resurrection (Lk 20:27-40). I do not believe that Jesus deals with the question of marriage and sex in the new heavens and earth. His chastisement aims directly at their Sadducean lineage, which is unprepared for the age of resurrection. In essence, Jesus puts a different question to his provocateurs: "Are you even prepared for the 'age of resurrection'?" (see Lk 20:35). The implied answer to his rhetorical point is, "No, you are not prepared."

How is Jesus' response to the Sadducees actually an attempt to refocus them away from the issue of marriage and on their own status at the resurrection? The Sadducees—those who deny the resurrection (Lk 20:27)—approach Jesus with a conundrum. They want him to define the nature of resurrection with respect to multiple marriages in this life. Borrowing a scenario from the apocryphal book of Tobit, they ask him, if a woman is married and widowed multiple times, whose wife is she in the resurrection? Jesus' answer seems to abolish the marital relationship in the age of resurrection. "The sons of this age *marry and are given in marriage*, but those who are considered worthy to attain that age and to the resurrection from the dead neither *marry nor are given in marriage*" (Lk 20:34-35).

Many have taken this as Jesus' definitive annihilation of the institution of marriage, an institution established from creation itself.[13] Maybe Jesus is

---

[13]See William Loader, *The New Testament on Sexuality* (Grand Rapids, MI: Eerdmans, 2012), 434n11.

describing a sexless or marriage-less eschaton, or both. But let us consider how Luke constructs his conceptual world with this language about marriage. Then, we can see that Luke's Jesus is using marriage talk in a technical way to make a point about obstinacy rather than abstinence.

First, Luke has already put the reader on notice about discerning the times. In Luke 12, Jesus chastens his hearers because they are able to perceive invisible meteorology favorable to their agrarian needs but fail to interpret "the present time" (Lk 12:56). Specifically, they can discern the possibility of nourishing rains for their parched fields from a mere cloud on the horizon. Using the same discernment, they know the likelihood of heat from a wind blowing north out of Egypt that will scorch their crops. Jesus' anger indicates that they do not employ their skills of interpretation equally, that is, in discerning the signs of the kingdom now present. The problem of going about one's business and not noticing the signs of the time appears to be a negative theme in Luke. Hence, Jesus calls them hypocrites; they interpret the weather but not the times.

Second, Jesus' phrase "marrying and giving in marriage" appears carefully chosen and should jog the reader's memory. That same phrase, "marrying and giving in marriage," appears in precisely one other place in Luke's Gospel: Jesus' teaching on his return and judgment of humankind (i.e., the age of resurrection, Lk 17:27; parallels in Mt 22:30; 24:38; Mk 12:25). In that context, Jesus uses the odd phrase—"marrying and giving in marriage"—not to talk about marriage but to describe the unaware people in the days of Noah and the destruction of Sodom and Gomorrah. What were they doing to deserve divine annihilation?

[People in the days of Noah] were
eating and drinking
marrying and being given in marriage (see Lk 17:27)

[People in Sodom and Gomorrah] were
eating and drinking
buying and selling
planting and building (see Lk 17:28)

Considering what we know about Sodom and Gomorrah, we might be surprised to find the crimes Jesus lists were not crimes at all. In fact, Jeremiah once advised the Babylonian exiles to do these exact things almost six centuries prior: build, plant, marry, and give in marriage (Jer 29:5-7).

In other words, the phrase "marrying and giving in marriage" cannot mean in Luke what it plainly might otherwise mean. Even more, this phrase does not seem to refer to marriage any more than Sodomite "buying and selling" and "planting and building" refer to those actual activities. Rather, this phrase signifies the mentality of those who are blinded to the coming judgment by going about their business and ignoring the obvious signs, or something to that effect.

This also explains the cutting question latent in how Jesus describes them as "those who are considered worthy to attain to that age," intimating that the Sadducees are not worthy (Lk 20:35). What begins as a lesson in the metaphysics of the permanence of marriage ends with a rebuke that equated the Sadducees with the people who died by the hand of God in the flood and in Sodom.

Whether or not we believe that Luke is using "marrying and giving in marriage" as a technical phrase, or whether we think Luke's Jesus is not talking about marriage, the larger context of sex should be kept in mind here. Sex is somehow connected to the image of God (Gen 1:27). The marriage of male and female for generation is the first blessing from God on humans (Gen 1:28).

Unlike the tabernacle, the laws of Moses, the prophetic rebukes, and so on, marriage is not an accommodation to something gone wrong. Like the Sabbath rest of God in creation, marriage and generation are part of the fabric of creation before anything goes wrong. They are ordering principles in the universe. In the conceptual world of the biblical authors, one singular rebuke by Jesus to folks who do not believe anything about which they are asking seems like egregiously tenuous grounds to deny the persistence of a fundamental aspect of creation: namely, marriage. Is marriage modified in the eschaton? I would imagine so. How so? I do not feel gutsy enough to speculate in print, but I trust that it is a good-making transformation.

Now we can see a Darwin-like tree of life conceptually extending into the New Testament literature. We no longer see an emphasis on Israel's generation that fruitfully expands the flourishing justice of the Torah from Israel into the world. Instead, the pruning of this tree happens through fealty to the new messianic prophet announced in Jesus (e.g., Acts 2:22). In fact, Jesus uses this exact analogy to indicate the proliferation of some and the extinction of others in the eschaton:

> I am the true vine, and my Father is the vinegrower. He removes every branch in me that bears no fruit. *Every branch that bears fruit he prunes* to make it bear more fruit. You have already been cleansed by the word that I have spoken to you. Abide in me as I abide in you. Just as *the branch cannot bear fruit by itself unless it abides in the vine,* neither can you unless you abide in me. I am the vine; you are the branches. Those who abide in me and I in them bear much fruit, because apart from me you can do nothing. *Whoever does not abide in me is thrown away like a branch and withers; such branches are gathered, thrown into the fire, and burned.* (Jn 15:1-6 NRSVUE)

John's Jesus pictures the people of God as lineages of a vine, some grafted in and others cut off and made extinct. The pruning mechanism in this analogy is not natural selection but God himself. He equates the pruning to "the days of Noah" and "the days of Lot" or those who are unworthy to "attain to that age and to the resurrection" (Lk 17:26-28; 20:35).

I do not want to suggest that Darwin borrows a uniquely Hebraic notion of treelike generation and pruning from Scripture, but only to notice their similarities for the sake of comparing them apples to apples.

CONCLUSIONS

After all this discussion of sexual and asexual reproduction, forced and coerced copulation, and monogamous coupling beyond sexual reproduction, where do these two conceptual worlds land? The goal of generation seems to be an obvious candidate for bisecting these two conceptual worlds.

Throughout Darwin's works and the proposals of some evolutionary scientists, the goal of sex is propagation of genetic material in a struggle for life, whether cooperative or competitive. The genes of those most amenable to the habitat (genetic influence) and those most amenable to the females' preferences (cultural influence) win. The sexual means of propagation tend to follow the biological makeup of the species: force and coercion where males can dominate females; participation in mating rituals where they cannot. The natural history of hominin propagation might possibly be a history of sexual assault—forced copulation—on this view.

The biblical texts appear to have a singular generational interest. They are anything but prudish about portraying various and deleterious sex practices. However, there is a telos to reproduction beyond spawning a particular genetic lineage: to monogamously reproduce, stay, play, and raise offspring who will embody the justice of the Torah under the fear of Yahweh within a nation welcoming non-Hebrews to do the same.

Israelite generation metaphors also focus on propagating justice and equanimity as much as any genetic lineage. And generation is not depicted as the only fundamental good.[14] Foreigners and nonreproductive eunuchs are part of the plan to expand Israel's justice into the cosmos (e.g., Is 56). Hence, the shift of focus off the normative sexual generation of Israel onto sending and making disciples from among the nations should surprise few readers of the Old Testament.

Finally, what are some points of continuity and potential conflict between the evolutionary sciences and the biblical emphases on generation? First, if it were the consensus that all sex evolved from asexual reproduction (currently debated), then against the reasoning of the biblical creation accounts, there might have been a time when reproduction was not sexually differentiated. Second, human sex is uniquely and cooperatively coupled in the biblical creation accounts in Genesis. For hominins, there was a time and place where we can safely presume the biblical authors believed that forced copulation was not practiced. This is not presumed or entertained in the evolutionary accounts.

Third, the sexual ethics based in the Hebrew creation accounts and further refined in the Torah would appear entirely disadvantageous to most versions of natural selection. This would make the biblical sexual ethics nonselective from the perspective of selection if the goal were spreading one's genes as broadly as possible.

Fourth, as a possible point of contact between evolutionary and biblical origin stories, extinction of one's name and genetic line is an ever-present fear across the biblical literature and leads to cooperative and coerced copulation in Scripture. Similar pressures might be in mind under evolutionary psychological models of sexual propagation.

---

[14]E.g., the propagation of incorrigibly wicked people is met by the divine warrior in the conquest of Canaan and later the destruction of northern Israel by Assyria.

Fifth, Jesus calls his peers back to the same goals of generation in the Torah, reminding them about divinely executed extinction events, such as the flood and Sodom. He also employs normal generational sex and agricultural fruitfulness, among others, as broad metaphors for the nature of the expanding empire of God. Moreover, Jesus' recurring references to an extinction episode in Genesis are precisely to the episode that involved nonprocreative forced copulation and sexual violence (i.e., Sodom and Gomorrah).

Sixth, only the most generous and ideal version of a cooperative natural selection model could foster exclusive coupling like the kind valorized in the biblical texts. Outside that moral version of cooperative selection, it is difficult to avoid the problem of grounding hominin development in a long history of forced copulation, including all human history.

THROUGH TWO GLASSES, DARKLY

FINALLY, I WILL ATTEMPT to wrap this up and tie it together as best I can. This requires a certain kind of fantasy life and a thriving imagination from us.

Just as archaeologists ask us to imagine an ancient homestead emerging from a faint outline of stones in a ruin, so too the biblical authors demand our imaginations. They cast a vision of an indistinct but discernible other way that the physical world could exist. As it once was, is no longer, yet evermore could be, they tell us of a renewed heavens and earth. Isaiah declares that new world as the source of hope for Israel and Gentiles alike (Is 56; 66). Paul refers to these as mysteries (1 Cor 15:51). John sees visions of it (Rev 21–22).

Just as important as conceiving of the world renewed, we must wrestle with what precisely needs renewing. Good thinking that incorporates the evolutionary sciences might still struggle to reconcile this imaginative aspect to the Hebrew intellectual world, a world that funds the New Testament's imagination. The metaphysical ruins of this cosmos will one day be renewed by reorientation. The goodness of creation in the beginning is obscured to us by its present deformities. But the entirety of the good news is predicated on this metaphysical misorientation that only Jesus can reorient in his return. Is there a way to reconcile entirely the Hebrew intellectual world to the present evolutionist accounts, theistic or otherwise? I am now less sure whether they can be reckoned without remainder.

In the end, the evolutionary sciences and the biblical authors might have two different metaphysical and political schemes: one free of ethical considerations until they later emerge in culture through evolution, the other intertwining metaphysics with ethics and divine politics. For the Hebraic thinking of the biblical authors, human action in the world metaphysically reconfigures the cosmos.

THE REQUISITE PHILOSOPHICAL IMAGINATION

We all know someone subject to the arc of addiction or who has found freedom from it. Maybe it was us. Let us imagine a young adult who once embodied the seemingly infinite possibilities for prospering. She was humble and curious, brave and sincere. Her introduction to illicit drugs came during a brief phase of instability and took her down the rabbit hole of addiction. After just a few months, her youthful body had been co-opted. Her entire being bent and habituated her endocrine system and neurotransmitter pathways toward one laser-focused goal: just one more hit.

Networks of anxiety that could have helped steer her away from such traps of addiction now surge within her at the thought of the next dose. Everything in her that was meant for flourishing now has tapered her faculties to the well-timed and appropriate amount of the needed chemical. In the cruelest irony, the drug that once offered euphoria now demands everything from her desires and physiology only to give a fleeting sense of calm and relief. Then the pangs soon arrive, yet again, to consume her for the next spin in the cycle of addiction.

Eventually, she will rehabilitate and stop the cycle. Through several years of struggle and help from family and other social supports, she will emerge as a confident former addict. She has left addiction but gained a profound wisdom for the proliferation of addictive forms of life, forms not only found in illicit drug use. After training and focus, she goes on to coach and write incisively about our addiction-prone world, starting nonprofits to address local addictions in various communities.

Though this analogy breaks at several points, it contains enough real-life experiences to help us think about the imaginative task the biblical authors lay before us. For them, wisdom and discernment are part of the intellectual task. This is true for us today. A surgeon must be able to imagine a patient's life free of present symptoms. Imagination is not a fluffy, ethereal disconnectedness from reality as we often romantically paint it.

The biblical authors require this kind of intellectual imagination not only to understand the future eschaton or the past creation accounts but to understand our present cosmos. Just as the former addict has trouble remembering what she even thought or felt prior to her addictions, she is equally unrecognizable today to those who only knew her in the depths of her struggles.

I want to suggest that the biblical authors viewed their own present world as if it were in the depths of addiction. The cosmos is currently warped, and every good system has been commandeered toward ill ends. Left without God, we bend every institution toward the next hit, as it were. So, the biblical frameworks and concepts of political structures focus on their misuse, corruption, and potential harm against the vulnerable. The legal reasoning of the Old and New Testaments aims at desires gone awry and communities that will create exploitative practices under the gravitational pull of our corruptions. The dirt itself does not do what it ought to do because of the loitering power of this conceptually thick malady that they call sin.

Inside the throes of addiction, we can no longer imagine what it was like to not be addicted. It becomes so normal that the anticreational rituals of addiction—needles, fire, pills, and smoke—now feel natural in our lungs and veins. We also cannot imagine what it will be like to no longer be addicted, even if we desperately long for it.

While we get clear answers to neither "what it was like" nor "what it will be like," the biblical authors give us enough to command our imaginations and check whether they can be reckoned with what we know about the cosmos today.

This chapter aims to offer some modest imaginative constructions of the *tikkun olam*—the world to come. What can we say according to what we must say? From within the theology of the biblical literature itself, I suggest that Paul's through-a-glass-darkly metaphor goes both ways: back to Eden and ahead to the eschaton. Even Paul's analogy of looking through a glass darkly (or mirror dimly) requires both continuity and discontinuity. It is continuous in that the glass shows us only a distortion of that which we know today. The discontinuity appears more exaggerated when we imagine that the distortion will disappear in the uniting of the heavens and earth.[1]

When Sir John Polkinghorne, a Cambridge physicist turned theologian, wrestled with how to make the Trinity fit with findings from science, his imagination came to the fore as the most responsible tool for working it out. Because the biblical story line aims us into a new heavens and earth, he needed to imagine how that completed the scientific narrative as well. He concludes, "Transformation that will overcome futility and bring about God's final purposes can be neither just a repetition of this world nor simply an apocalyptic wiping clean of the cosmic slate."[2]

The futility of this world, according to Polkinghorne, can only be overcome by imagining "a new kind of 'matter' endowed with internal organizing principles of such power as permanently to overcome any tendency to disorder."[3] The world to come must have sufficient continuity with our world, physically speaking. At the same time, the age beyond resurrection must have sufficient discontinuity as not to metaphysically regress to futility akin to the cursed dirt, wombs, desires, and cosmos lamented in Genesis 3. Assessing the same problem, Richard Middleton likewise affirms that the age of resurrection is presently "hidden or obscured by the continuing power of sin."[4]

TWO INTELLECTUAL WORLDS

Comparing the intellectual worlds of evolutionary science with those of the biblical authors, a trend materializes. Both worlds interweave reactions to scarcity, fittedness to place, and sexual propagation into their origin stories. Both origin stories intend to explain the present. Both cast scarcity as a fuel for violence or opportunity for cooperation, sex as a tool of genetic exploitation, and creatures either fitting or mis-fitting their environment. However, the biblical authors seem to place the provision of resources, the mutuality of sex, and the need for environmental fit into the realm of divine-human

---

[1]"The apocalyptic pattern assures us that even though we may not see the promised kingdom clearly in the world today—it is in many ways hidden or obscured by the continuing power of sin—God guarantees the final success of this kingdom." J. Richard Middleton, *A New Heaven and a New Earth: Reclaiming Biblical Eschatology* (Grand Rapids, MI: Baker, 2014), 221.
[2]John Polkinghorne, *Science and the Trinity: The Christian Encounter with Reality* (New Haven, CT: Yale University Press, 2004), 153.
[3]Polkinghorne, *Science and the Trinity*, 164.
[4]Middleton, *A New Heaven a New Earth*, 221.

relations. More specifically, the entire discourse of biological diversity is rooted in God's total jurisdiction over creation and humanity's response to his reign.

The biblical authors sometimes but not always subjugate Mother Nature's problematic brutality under God's judgments—sometimes (e.g., Ex 7–14; 2 Sam 21:1; Ps 105:16; Jer 11:22; etc.). To a lesser yet significant degree, they file it under the disjointing of the cosmos by humanity's response to God (e.g., Gen 12:10; Neh 5:3; etc.). In contrast, an evolutionary account of nature might portray the same brutality as the lathe of genetic shaping.

I also sound a warning to the theological tendency to secure the goodness of God against all of nature's cruelty by naive appeals to what we take to be true propositions in Scripture. The "image of God" statement in Genesis 1 often forms loose objections put to theistic evolution. As the *imago Dei* comes to the fore, we wonder what a human is and when hominins became the image of God in the evolutionary processes. Of course, humans are depicted as created in the "image of God, male and female," but the *imago Dei* theme is rarely explored openly by the biblical authors—much less the *imago Dei* as sexually differentiated. The surprising rarity of the image of God throughout biblical literature does not necessarily diminish its impact on their conceptual understanding of humanity. Its near absence in Scripture is striking mostly because of its near-universal appearance in later Christian theology. But what impact has that concept of *humanitas qua imago Dei* had?

Merely viewing humans as created in God's image in no way guarantees that such a view will lead to the humane treatment of others.[5] Recall that seventeenth- and eighteenth-century proslavery voices fully affirmed that Africans and indigenous Americans bore the image of God. Thus, they concluded, because of their inferior environment and savage incivility, they were appropriate objects of enslavement. We should be under no illusion that a simple view of humans made in the image of God creates an adequate ethical theology and practice. It alone does not do sufficient work.

---

[5]Livingstone shows that even those in the nineteenth century who fought to regard indigenous American tribes as also created *imago Dei* went on to advocate for their enslavement without any sense of irony. In other words, singular divine origin of humanity still left plenty of room for enslavement, practically speaking, and for principled views of moral degeneracy of certain "primitive races." David N. Livingstone, *Adam's Ancestors: Race, Religion, and the Politics of Human Origins* (Baltimore: Johns Hopkins University Press, 2008), 64-68, 99, 122.

However, the conceptual schema of image of God, male and female, is developed across Scripture in terms of the warped world in which we now find ourselves. Jamie Grant proffers the anthropological vision of biblical poets that relies on a metaphysical view of the cosmos-now-crooked. In the misshapen world depicted throughout Scripture, humanity is not best understood by us as *imago Dei* but rather "frustrated man" (or less eloquently: *hominem pudore cooperti*).

Why is humanity frustrated? Grant gives three reasons from the wisdom tradition: "the imponderability of God's sovereignty, human limitedness and the inevitability of death."[6] The wisdom tradition of Israel promotes a counterintuitive demand for generations that spread humanity out into a progressively violent world with scarce resources. In this spread away from Eden, humans struggle to understand God's regency over the environment and over their reproduction. Thus, they contend with finitude that ends in death. Biblical authors hint at this frustration throughout and examine it floridly by means of apocalyptic literature such as Daniel, sections of Paul, Revelation, and so on.

God will eventually reorient the world to its intended metaphysical state. In this renewed world, the complete jurisdiction of God no will longer frustrate humanity. The heavens and earth will be united under the political regency of God. Human finitude will remain, but death and its associated frustrations will be somehow outstripped by a new nature, an environment to which only some people and things are properly fitted.

This renewed nature must be imagined from its continuity with the so-called natural world we encounter today: "a new kind of 'matter' endowed with internal organizing principles of such power."[7]

THE APE, US, AND CHRIST THE ÜBERMENSCH

A persistent question between the conceptual worlds of evolutionary and biblical metaphysics remains: Can we have a renewed heavens and earth without a necessary and persistent metaphysical connection to the

---

[6]Jamie A. Grant, "'What Is Man?' A Wisdom Anthropology," in *Anthropology and New Testament Theology*, ed. Jason Maston and Benjamin Reynolds, Library of New Testament Studies 529 (London: Bloomsbury, 2018), 5-25, here 7.
[7]Polkinghorne, *Science and the Trinity*, 164.

cosmos-turned-crooked? In other words, do the biblical authors want us to imagine a good world now corrupted and yet to be righted? Or, do they imagine a ruined cosmos soon to be destroyed and created de novo?

How seriously should we take the biblical depiction of Eden's garden as that which God intended and everything eastward as cancerous and crooked? We have indications of Genesis's and Paul's renewable world from various findings of science—lifespan genetics, psychedelic drug research, and more—but no clear picture of what the unbent cosmos was like without a biblically guided imagination.[8] I will return to some of these indications in a moment, but they are only that: possible indications that our presently listing cosmos can be metaphysically righted.

We can conceive but not necessarily comprehend our future bodies and a future cosmos oriented rightly at every granularity. Equally, we cannot comprehend humanity's estate in Eden but through a looking glass darkly. Theistic evolutionary accounts, in particular, must also wrestle with the fact that the biblical authors showed concern for all the evolutionary pressures we care about today, yet enshroud its details behind fogged glass. And though we cannot comprehend the physics and physiology of the age to come (not that anyone comprehends physics today), there should be indications of the eschaton that help us to conceive of such bodies and relations.[9] According to Paul, indications are all we get, but we can trust these as sufficient to understand the present (1 Cor 13:12).

The biblical authors present humanity, after Eden, like Nietzsche's man at present who stands holding a rope in each hand. One end of the rope stretches to the hand of the ape, representing hominins past. At the other end is the "overcoming man," the *Übermensch*. Current man stands in the middle, tugged on each side by the primitive forces of his past and conspiracies of what he could be in a future, evolved state.

---

[8]We could potentially add cancer research, placebo studies on ritual and self-healing, and even utopian pipe dreams in the realm of political order (played out in cooperative and feminist critiques of neo-Darwinian evolution) to the list of suggestive indications of what is on the other side of Paul's dark glass.

[9]Sean Carroll, "Even Physicists Don't Understand Quantum Mechanics: Worse, They Don't Seem to Want to Understand It," *New York Times*, September 7, 2019, www.nytimes.com/2019/09/07/opinion/sunday/quantum-physics.html/.

In the biblical literature, we stand between two ages of the cosmos—currently mired in the twisted span in between them. As we look back to the biblical depiction of the garden, we look through glass dimly. So too with the age of resurrection, comprehended narrowly through prisms such as Jesus' actual resurrection. Those suggestive refractions are only hinted at through instances such as Lazarus's resuscitation. Unlike Nietzsche's program for overcoming our present values, Christ is depicted as our *Übermensch*, overcoming all metaphysical and political powers.

METAPHYSICAL CHANGES TO NATURE

What can we affirm about our so-called natural history? Christians must land somewhere about the nature of the natural world. Namely, is nature constituted as Michael Lloyd claims, "riddled with pain, death, disease, and predation? And a moment's reflection will further reveal that these are not incidental to creation but appear to be built into its very fabric."[10]

Lloyd is correct to say that this is what we see when we look around. But that only highlights the problem with the assertion: the assumption that what we see is the way it has always been and will be. For him and others like him, there was no time when the cosmos was not fully or potentially riddled with "pain, death, disease, and predation." For him, these reveal the nature of nature itself.

In the biblical story line, God metaphysically disorients the world toward scarcity, and hence the pains accompanying it, through the curses on woman and man. Though death appears possible at some point in the world of Eden (Gen 3:22), it seems to be staved off by the presence of God in the garden.

East of Eden, this contorted situation creates the conditions for Yahweh to later promise a metaphysical reorientation of the weather, soil, wombs, and political states in favor of a flourishing and justice-filled Israel. So, even in the Torah, we find the reasons for anticipating a metaphysical reorientation of death in the form of resurrection.[11]

---

[10]Michael Lloyd, "Theodicy, Fall, and Adam," in *Finding Ourselves After Darwin: Conversations on the Image of God, Original Sin, and the Problem of Evil*, ed. Stanley P. Rosenberg et al. (Grand Rapids, MI: Baker Academic, 2018), 244-61, here 244.

[11]Jon D. Levenson, *Resurrection and the Restoration of Israel: The Ultimate Victory of the God of Life* (New Haven, CT: Yale University Press, 2006), 23-34.

But the indications of resurrection throughout the Hebrew Scriptures and its proclamation across the New Testament raise a substantial question: Is the resurrected body the same wrecked body brought up from the grave into the same cursed cosmos? The answer later clarifies that the resurrected body and the united heavens and earth fit each other because they have been reoriented to fit, even if they appear veiled to us now.

Metaphysics and not mere morality. I have been using the word *metaphysical* to refer to the misorientation and reorientation of the cosmos. Unfortunately, the term *metaphysical* does not mean the same thing to everyone. According to Peter Van Inwagen and Meghan Sullivan in their entry on metaphysics in the Stanford Encyclopedia of Philosophy, "The word 'metaphysics' is notoriously hard to define."[12] One simplistic attempt to define *metaphysical* is "anything to do with the nature of relations in the physical realm." It can also refer to the study of "being as such," which will not mean much to nonphilosophers. In the Hebrew intellectual world, when we speak of the cursed distortions of the physical world, then we are talking about metaphysical changes to the cosmos—the way the stuff of the cosmos is constituted and oriented, even in relation to human activity.

Al Wolters argues that the best lens for clarifying what the biblical authors seem to suppose is the structure-direction difference.[13] The structure can be regarded one way—fundamentally good—yet the orientation of the structure can and must be regarded separately for the concept of redemption to function within the Hebrew Bible and New Testament.

Cancer illustrates the analogy for structure/direction well. Purportedly, a cancer cell begins as a human body cell (structure) that turns against the body and convinces other cells to do the same (direction). The structure is good; the misorientation is ruinous. The same goes for our addict mentioned at the outset of this chapter. She was a good human in structure, but entirely usurped and disoriented by her addiction. Disorientation entails a potential to be reoriented to a life free of addiction. Thus, redemption, as a metaphor, flows from this structure-direction paradigm.

---

[12]Peter Van Inwagen and Meghan Sullivan, "Metaphysics," in The Stanford Encyclopedia of Philosophy, Winter 2021 ed., ed. Edward N. Zalta, https://plato.stanford.edu/archives/win2021/entries/metaphysics/.

[13]Albert M. Wolters, *Creation Regained: Biblical Basics for a Reformational Worldview*, 2nd ed. (Grand Rapids, MI: Eerdmans, 2005).

A metaphysical misorientation of physical processes helps us to make sense of that which the biblical texts present, a good structure turned awry. Notably, the dirt turns against the dirtling. It is not merely that humans turn against the earth in an ecological view of humanity. I will remind the reader here of Deuteronomy 28 and its argument for metaphysical changes of blessing and curse already rehearsed in chapter 12. All of creation could be fruitful by following Yahweh's instruction. But rejections of Yahweh's guidance would turn the same facets of reality against Israel. Their wombs, fields, flocks, and more would be metaphysically misoriented by Yahweh in relation to their moral lives as a community.

Why did the dirtling die? In case we thought this was describing some kind of naturalized wisdom that leads to prosperity, the curses quickly convince us otherwise. No natural process—free of divine involvement—singles out a people group for either fruitfulness or destruction according to their communal practices of justice and mercy. It is the specificity of the curses that rules out anything but a metaphysical reorientation of earth and empire to Israel.

So too with the dirtling and the dirt, the woman and childbirth, the serpent and humanity in Genesis 3. What could have been a fruitful relationship is now metaphysically redirected for dearth, toil, and pain.

Over the course of the Torah and Gospels, we come to understand that the metaphysical nature of these curses is what makes all of creation simultaneously cursable yet redeemable in the thinking of at least some biblical authors. Cursing disorients creation's situation, blessing reorients. But these matters of reorientation only matter if the affairs of humanity also have a metaphysical impact on creation itself.

Humanity alone is tasked with manifesting God's care, justice, and right relations on earth. Humanity alone ruins the earth. And God holds humanity alone responsible for its disorientation and systemic injustice (see the Torah's and Jesus' threats of judgment and exile). Though animals can be morally culpable for murder, humanity's actions uniquely affect the metaphysical orientation of the cosmos and are conjoined into its reorientation in the renewal of the heavens and earth.

If this all tracks truly, then the exile of humanity from God's presence in Eden and God's ensuing retreat into the heavens creates metaphysical, and hence physical, disorientation. Potential death now actualizes as the first couple leaves Yahweh's life-animating presence and the disorientation of the physical cosmos creates the gravitational pull of the dirtling's life back down, ever into the dirt.

In Genesis 3, God's threat—"in the day you eat . . . dyingly you shall die" (Gen 3:3 my translation)—accurately refers to one particular event juxtaposed against eating from the tree of life: the couple's exile from Yahweh's presence. Exile that begets decay and disorientation of the physical world emerges as the long-form death assumed in Genesis 3.

Under most definitions, everything said above requires us to imagine some kind of metaphysical turning within and between the structures of creation because mere entropy—irreversible fragmenting—does not sufficiently address the concepts being developed and referred to across the biblical texts.

Is Eden an island of rightly oriented earth? Some might wonder whether the biblical texts focus only on what happens in the garden while assuming something entirely different outside Eden. Hence, the world had long experienced "pain, death, disease, and predation," and this is precisely the world that the man and the woman step out into.

Are physical decay and predation normative outside the garden? I am unsure. Walter Moberly addresses a similar silence in the stories of Genesis 4—specifically addressing the ancient reader's question of where Cain's wife came from. Instructively, he says, "The details of the Cain and Abel story show that the Bible is aware of a larger human history that it chooses not to tell."[14] Both are important: that the text reveals its awareness of "a larger human history" and that the narrator stays silent on that history. The biblical authors are not hiding anything. They are ignoring everything west of Eden in the untold prelude to the garden.

The text is interested in telling a particular history. The creation of beasts, birds, and fishes alongside the sun and moon suggests that the garden was metaphysically normal in creation. In the conceptual world of the biblical authors, the garden does not appear (to me) to be a metaphysically unique playground for the dirtling and the woman while decay and corruption

---

[14]R. W. L. Moberly, *The Theology of the Book of Genesis* (New York: Cambridge University Press, 2010), 26-27.

infect the rest of creation east of Eden. The sun, moon, and stars appear to be part of the rightly oriented universe—"let them be for signs and for seasons, and for days and years" (Gen 1:14)—and they are nowhere near Eden. They participate in the system of Eden, as do the flora and fauna.

Unfortunately for the curious, the narrator remains silent on what is happening outside the garden, and our only glimpse of the exterior world occurs after metaphysically corrupting curses have already taken effect.

But inside the garden at Eden, we do not even get details, suggestions, or intimations of that larger human history, or of the histories of flora and fauna outside the garden. What we get is a story of humanity's shifting loyalties that ends in exile with metaphysical consequences for the cosmos.

This dying-in-the-day-that-you-eat story either makes God into a day-stretcher—the man did not die for another three hundred thousand-plus days (Gen 5:5)—or indicates that the exile from Eden is itself a form of Sheol's reach into the land of the living.[15] If the eventuality of death through deterioration is what the biblical authors assume in Genesis 3, then it is physical death that begins in that day, not the so-called spiritual death (not that these are mutually exclusive, depending on what we mean by *spiritual*).

In brief: it is not the case in Genesis 3 (or Deut 28) that humanity's relation to the earth alone is changed but also the earth's relation to humanity. The metaphysics have changed. According to some biblical authors, there was a time when some place in creation was not oriented toward scarcity, pain, and death.

If the above is correct, even if partially, then no study of the present and corrupted world can domesticate and comprehend the fantastical worlds of Eden and the eschaton. From the perspective of the biblical literature, the present fog of our world is prohibitive for certain leaps of comprehension. However, there ought to be indications of what is beyond the darkened glass.

INDICATIONS OF OUR RENEWED NATURE: PSYCHEDELICS AND LONG LIFE

This will be too brief but is worthy of consideration. If our cosmos is dissimilar yet continuous with creation and the eschaton, then we should

---

[15]Levenson compellingly argues that *she'ol* ("the grave") is sometimes portrayed as beginning in life, not merely a metaphor for death (*Resurrection and the Restoration of Israel*, 37-39).

expect to see physical indications of the possibility of this coming meta-physical reorientation. I will speculatively present two possible signals from our present cosmos to the creational past and the future depicted in the biblical literature. If the presently defunct cosmos creates tensions resolved by evolutionary models, then we can also think in terms of resolving some of those tensions with our biblically shaped imagination. We are attempting to imagine some indications of a future metaphysical reorientation that might align with the fantastical images portrayed in Scripture.

Psychedelic brains. This is in no way an endorsement for using psychedelics (quite the opposite). In the new heavens and earth, scarcity is resolved by harmonious provision. Sexual generation based in fear of extinction is eclipsed by flourishing secured into the distant future. Environmental fit is felt, known, and a mechanism of human responsiveness to creatures, creation, and God.

Imagine a fantastical world in which finite humans persist for indefinite lengths of time; where we feel and think from a profound sense of connection to God, humanity, and all the structures of creation entirely bent into mutually beneficent directions; and where we bear no pangs of anxiety about diminishing resources. This imaginative space now has some empirically suggestive support. Let us consider two areas of potential: psychedelics and life extension.

Recent clinical research with hallucinogens has provided insights into humans beyond the way we have always experienced ourselves. Brains on psychedelics appear to bridge out of normal neural ruts and cross-talk with other areas of the brain. At the same time, hallucinogens tend to cause one's sense of self to recede. "The brain operates with greater flexibility and interconnectedness under hallucinogens," "implying that [subjects' brains] communicate more openly but, in doing so, lose some of their own individual 'identity.'"[16]

It is not just a recession of the self but also a sense of relation to other things and other folks. The subjects of psychedelic research had a strong sense of connectedness to nature and others.[17] This felt sense of connection

---

[16]"Hallucinogens have a disorganising influence on cortical activity which permits the brain to operate in a freer, less constrained manner than usual." Robin L. Carhartt-Harris et al., "How Do Hallucinogens Work on the Brain?," *Psychologist* 27, no. 9 (2014): 662-65.

[17]Michael Pollan, *How to Change Your Mind: What the New Science of Psychedelics Teaches Us About Consciousness, Dying, Addiction, Depression, and Transcendence* (New York: Penguin Books, 2018), 316.

to the environment even has measurable long-term effects. A person's past use of psychedelic drugs can be predictive of environmental actions later in the person's life well beyond their period of using hallucinogens.[18]

Imagine, even if only temporarily, becoming a low egocentric person with high cognitive flexibility and a profound sense of connection to everyone and everything around you. The effects of experiencing that version of ourselves might explain why even atheists are having experiences of transcendence on clinically guided psychedelic trips.[19] Such was the case with one cancer-stricken woman seeking to reconcile her anxieties about death: "Not only was the flood of love she experienced ineffably powerful, but it was unattributable to any individual or worldly cause, and so was purely gratuitous—a form of grace. So how to convey the magnitude of such a gift? 'God' might be the only word in the language big enough."[20] We should notice that it is not the psychedelic drugs that are operating across the brain. The actual drugs in the brain tissues do not appear to be making these neural connections happen at the site of each new neural connection. In other words, it is not as if the drug acts as a physical bridge between parts of the brain that do not usually talk to each other.

Rather, and this is the eschatological indication, hallucinogens activate the 5-HT2A receptor in the brain's cortex in a new way. The sense of nature connectedness, general openness, and neural flexibility of the brain is a biological reaction to this receptor interaction. To say it another way, the brain itself might be able to operate this way biologically if it were reoriented to do so, no hallucinogens needed.

Now we can imagine, even if we cannot fully grasp, how an entire community of such connected and relational persons might interact when we remove the anxieties of death and disease (e.g., Rev 21:3-4). This is the fantastical world that the biblical authors construct with God as king.

---

[18]Matthias Forstmann and Christina Sagioglou, "Lifetime Experience with (Classic) Psychedelics Predicts Pro-environmental Behavior Through an Increase in Nature Relatedness," *Journal of Psychopharmacology* 31, no. 8 (June 2017): 975-88.

[19]Alongside the significantly increased sense of nature connectedness and ego dissolution with psychedelics, subjects also express a significant increase in negative views toward authoritarian political structures. Taylor Lyons and Robin L. Carhart-Harris, "Increased Nature Relatedness and Decreased Authoritarian Political Views After Psilocybin for Treatment-Resistant Depression," *Journal of Psychopharmacology* 32, no. 7 (2018): 811-19.

[20]Pollan, *How to Change Your Mind*, 285.

Long life in our land. This is in no way an endorsement of life extension programs. Living well beyond 120 years old is no longer medically inconceivable. As a matter of genetic science, our lifespan has no clear limits apart from limitations imposed by our genes and the functions of cellular regulation by specialized proteins such as mTOR. Apart from our common experience of old age, a lifespan above one thousand years requires just as much scientific explanation as a lifespan above one hundred years, or fifty years, for that matter. Humans live to around a century when health is good and genetics favor them. But what if our cells' ability to reproduce were no longer limited?

The caps of our DNA, known as telomeres, appear to control how many times our cells can reproduce.[21] This means that our own DNA has already set how many times our liver cells or muscle cells reproduce. This limiting effectively sets our possible age limit when taken cumulatively.

But why do our cells quit reproducing and allow our bodies to decay unto death despite the fact that our environment, nourishment, and exercise remain the same? Why do our cells not continue to reproduce the same number of healthy cells forever and ever?

Though geneticists hoped that lengthening these telomeres could increase our lifespan, it has turned out to be more complicated.[22] Nevertheless, the fact that the orientation of our biological structures determines our length and quality of life is less complicated, even suggestive that there is a metaphysical orientation of the world in which it might not be the case.

We can now picture the possibility of communities of humans with bodies (including brains) metaphysically redirected in a way that favors their connectedness to each other, creation itself, and God.

But a new biology is not enough. After all, Genesis 4 immediately injects another form of death into human society. In other words, the hope of long-living bodies and hyperconnected brains mitigates but does not account for acts of murder and accidents. A political order is required. Indeed, this is what the New Testament authors repeatedly point to as signs of the kingdom of God, not merely a new *natural order*, having come in full.

---

[21] Brett Heidinger et al., "Telomere Length in Early Life Predicts Lifespan," *Proceedings of the National Academy of Sciences* 109, no. 5 (September 2017): 1743-48.
[22] Tad Friend, "The God Pill," *The New Yorker*, April 3, 2017.

The signs and wonders that establish the credibility of Jesus, and later the apostles, aim specifically at metaphysically redirecting all physical states of affairs. Jesus' healings do not make transhumans out of those who seek healing. They are not made better, stronger, faster than they were before. Rather, their broken and contorted bodies and tormented psyches were reoriented to a natural state—back to the way they were supposed to be. This signals to us what the physics of Eden and the eschaton might be like, where structures are continually directed beneficently toward flourishing with psyches (for lack of a better term) that respond in kind.

HYPOCRITICAL CHRISTIAN FANTASY WORLDS

Come with me, and you'll be, in a world of biblical imagination. For reasons I could only speculate, some Christians might have deep concerns about the idea that the cosmos, or just a portion of it, was metaphysically different from how we now experience it. It is too much of a cognitive leap to imagine creation and creatures both structurally and directionally oriented aright—that this present situation, as Cornelius Plantinga writes, is *Not the Way It's Supposed to Be*. For some, the Edenic account is too much a fictional fantasy and not enough critically real about the world.

While we all feel the entropy of the present crumbling cosmos as a center that cannot hold, Christians simultaneously entertain a robust fantasy about one particular Jew in the first century, born of a not-yet-sexually-active woman, who then died, rose from death, and is presently resurrected in a realm unexperienced by us, called "the heavens."[23] Most professing Christians realistically trust in a reoriented future of the new heavens and earth, a fantastically difficult-to-conceive-of universe where that same God-Jew named Jesus will judge, rule, and sustain a new world order we could all conceivably appreciate—long for, even.

So a rich fantasy life about a first-century Galilean that bends the laws of physics and a future resurrection that defies all current physiology is fair game for trusting. The Gospels and the eschaton are an unflinching

---

[23]By "fantasy," I mean only that the vision being cast requires our imagination because it is not a known reality to us. The term *fantasy* in this sense has no being on the truth of the matter, only what imaginative faculties are required to picture it.

Christian fantasy held with conviction by most Christians. Yet, an origins story that does not seem to fit with our current physics is a fantasy
gone too far?

The tragedy of this schismatic belief in resurrection but not in a metaphysically different past injects problems into the whole theological system.
For it is the biblical origins story that uniquely provides the conceptual and
metaphysical basis for the age of resurrection and the new heavens and earth.
Why is our imagination often limited to only the future heavens and earth?
Is it because there is not yet a fossil record of the eschaton to offend our
fantasies of the new heavens and earth?

Most of us are struggling with what to make of the already-found fossil
record, and that is a right and necessary struggle. The so-called natural
history of our planet has a lot of explaining yet to do. But I am less interested
in easing the burden of correctly interpreting the fossil record.

In this work, I am attempting to put the two intellectual worlds of
Hebrew thought and the evolutionary sciences in conversation. I will
outline the points of possible agreement and conflict below. However, it
now appears necessary to confront a problem that we might have with the
biblical story of the cosmos. It is us, not the biblical texts. Either we have
an imagination capable of trusting in fantastical states of nature past and
future, or we do not.

THE FANTASTICAL AND POLITICAL
END OF HUMANITY

Richard Bauckham's shrewd introduction to the book of Revelation boils the
apocalyptic book down to one question: "Who is Lord over the world?" And
what is John's unique answer, found across the book of Revelation?[24] "He
anticipates the eschatological crisis in which the issue will come to a head
and be resolved in God's ultimate triumph over all evil and his establishment
of his eternal kingdom."[25] By using the terms *lord* and *kingdom*, we should
not miss the technical meaning of these terms that have become pallidly

---

[24]Unique in literary mode and persuasion as compared to other Hellenistic Jewish apocalyptic
literature. See Richard Bauckham, *The Theology of the Book of Revelation* (New York: Cambridge
University Press, 1993), 9-12.
[25]Bauckham, *Theology of the Book*, 8-9.

spiritualized into stock Christian cliché today. Both *lord* and *kingdom* strike our ears as dead metaphors, though they were not for early Christians. Bauckham refers here to the New Testament's notion of a new political order insinuated by these political metaphors that also have God at the preeminent political position: the "lord" over a "kingdom."

In taking up a political metaphor insufficient to describe the totality of the eschaton, John and other New Testament authors imagine a total effectual change to the known world.[26] There will be no more death, which seems to mean that the age of resurrection was conceived of as a reattunement to the Eden-esque situation, for some. Survival, along with all its derived anxieties, is out.

But not all the resurrected are fit to the environment of the new heavens and earth. The wicked, the false, the detestable, and those who rebelled against God are considered misfits—not allowed (Jn 5:28-29; Rev 21:27).

Because the final metaphysical, personal, and political state of the world gets wrapped so heavily in the packaging of what we might call Edenic naturalism, it is difficult not to see Eden and the eschaton as parallel states of nature separated by a disoriented gap of history.

In that already inaugurated and now finally here state, food yields from an ever-fruiting tree of life, watered endlessly by a river springing from under the throne of the Governor of the universe (Rev 22:1-3). Light that scatters all fears radiates from the King himself (Rev 22:5). Xenophilia, as opposed to xenophobia, becomes naturalized in a city that welcomes foreign invasion without fear (Is 56:3-8; 66:18-23). Knowledge of God and his instruction will be ubiquitous between neighbors and brothers according to Jeremiah's vision of the renewed treaty (Jer 31:34).

Malachi and Luke wax fantastically about how family relationships will blossom: "He will turn the hearts of fathers to their children" (Mal 4:6; see Lk 1:17). As H. I. McDunnough envisions it in the Homeric classic *Raising Arizona* (a Levitical twentieth-century film by the Coen brothers):

And then I dreamed on,
into the future . . .

---

[26]Other metaphors will fill in some of the gaps. For example, the new heavens and earth will be like a bride and groom uniting, signaling a creation story of a man alone arriving at a quiescent state only once partnered with a woman.

where all parents are strong and wise and capable,
and all the children are happy and beloved.[27]

These eschatological images might intend to describe future historical realities to the first-century Jewish Christian. Or, Eden-like, they might mean to evoke an ideal circumstance fit for a first-century Jewish audience in a Roman-controlled Asia Minor. Either way, these fantastical images rely on one central fact in their return to Edenic nature: God's political rule ensures the natural state of affairs, not a set of so-called natural forces or laws.

Reading across Scripture, God's political power has always entailed his metaphysical power over creation, which includes sociopolitical affairs in history. Moses, after crossing the sea, sings of God on this same topic (Ex 15). Likewise, Hannah's song celebrates how the reach of God's jurisdiction goes deep into matters of life/death, poverty/riches, honor/shame, political powers, and all the way down into Sheol and the foundations of the earth itself (1 Sam 2:6-8). Paul later directly connects Jesus' metaphysical power over the universe and his political supremacy: "for in him all things in heaven and on earth were created, things visible and invisible, whether thrones or dominions or rulers or powers—all things have been created through him and for him" (Col 1:16 NRSVUE).

Hence, two primary manifestations of what Jesus refers to as "the mysteries of the kingdom of God" frequently attend his words and actions. First, Jesus' metaphysical powers restore people and creation in a glimpse of the world to come. He heals by naturalizing them to the way they were supposed to be, not a superhuman or transhuman state. Jesus frees humans of disease, orthopedically rights them, resuscitates the dead, and so on. In creation, Jesus quiets storms, multiplies food, makes water into a festal drink, and wrangles fish for income, with the notable exception of the cursed fig tree, which signals Israel's partially unfit state for this coming kingdom. These acts all trace trajectories of what nature in the world to come might actually be like when God takes over all regnal reins.

It now seems impossible to fully comprehend the "natural world of Eden" or the eschaton. The biblical authors give us hints and promises, and show us

---

[27]"Levitical" in that the film was written and directed by the Coen brothers. *Raising Arizona*, directed by Joel Coen and Ethan Coen (Beverly Hills, CA: Twentieth Century Fox Home Entertainment, 1987).

Jesus' renewals of nature in the locale of Galilee and Judea. And all of these appear to reflect what the prophets saw and advocated to Israel as the goal. These point insistently at something real through darkened glass, both backward and forward. The same fantastical imagination of an Eden unknown to us is required to talk about creation and humanity's history within it—a fantasy equally difficult for ancient Israelites to comprehend as for us today.

This fantasy requires that we see reality itself as fractured, but with the internal ability to be restored by reorientation. It is not the fantasy of Platonism, which imagines a true world in the heavens of which our experience is only deceptive, fleeting, and impressionistic. Nor is it the fantasy of Brahminism and Buddhism, which ask us to trust that none of our experience can be true or real. The intellectual world of the Bible advocates neither fleeing this material world for the heavens nor denying its objective reality altogether.

This Hebraic fantasy of the Old and New Testaments relies on our present experience of a disordered cosmos. Our experience of it is limited but veracious, even capable of indicating God's kingdom to us. Its dogma is not "believe this blindly," but rather, "trust and test your sense of the world."[28] Through trusting and testing, the Hebraic intellectual world supposes that one fantastical view, with broad explanatory power of the past, present, and future, will rise to the top.

A MORE PROFOUND CONFLICT THAN SCIENCE VERSUS RELIGION

For understandable reasons, most scientific thinking has effectively divorced itself from religious dogmas over the centuries. Even though our scientific communities are still plagued with the lingering dogmas of positivism, they produce practical understanding because of the necessity to check ideas against reality. Even so, the biblical concepts of communal knowledge, rationality, justification, and realism have contributed intellectually to the scientific enterprise as we know it today.[29]

---

[28]For more on the realistic epistemology of the Torah and Gospels, see Dru Johnson, *Epistemology and Biblical Theology: From the Pentateuch to Mark's Gospel* (New York: Routledge, 2017); or see the brief companion version for the nonspecialist, *Scripture's Knowing: A Companion to Biblical Epistemology* (Eugene, OR: Cascade, 2015).

[29]Dru Johnson, *Biblical Philosophy: A Hebraic Approach to the Old and New Testaments* (New York: Cambridge University Press, 2021), 237-307.

However, the stealthier dogmatic camels that have slipped into the tents of scientific communities have not gone unnoticed by critics. For instance, Tim Larsen argues that despite anthropologists' best efforts to free themselves from the odd and specific creation narratives of Genesis, they end up retaining a dogmatic view—not empirically based—on the unity of the human psyche: that all humans have a mind.[30]

While we might want to distance ourselves from dogmas to gain some clinical perspective, working scientists and the biblical authors alike depend on a set of commonly held beliefs, theories of knowledge, and models about human origins. Those might be worth spelling out, as we have discovered them here, before moving into points of agreement and conflict between them.

Commonly held paradigms in evolutionary science. No set of statements can uncontroversially capture what the evolutionary sciences suppose about hominin origins today. Yet, a series of paradigms and convictions surfaces repeatedly in the literature that I have engaged here. These occur with enough regularity to merit inclusion in a common description of "evolutionary science," even if each item would be debated for any given scientist.

First, after the emergence of nervous systems, there was *no* time when the earth was free of "pain, death, disease, and predation."[31] Though cooperation and competition could occur during resource scarcity and functionally affect natural selection, debates persist about when competition is deferred in favor of cooperation and vice versa. But this blanket statement rests on a more fundamental question: Is the "natural" disposition of the universe bent toward cooperation or competition? Though both clearly occur across the animal kingdom, which one gives way to the other, and under what circumstances?

Second, there was a time when sex might not have existed—when asexual propagation gave rise to sexed propagation (i.e., male and female). Or, more minimally, sexual reproduction did not have to emerge. Because sexed

---

[30]Timothy Larsen and Daniel J. King, "The Dependence of Sociocultural Anthropology on Theological Anthology," in *Theologically Engaged Anthropology*, ed. J. Derrick Lemons (New York: Oxford University Press, 2018), 54-65, here 64-65.

[31]Lloyd, "Theodicy, Fall, and Adam," 244. A recent book argues for the universal span of suffering from a Christian theological perspective. Bethany N. Sollereder, *God, Evolution, and Animal Suffering: Theodicy Without a Fall* (New York: Routledge, 2020).

differences in some animals split males and females into larger and smaller kinds, sexual propagation will include forced copulation (even if males are generally smaller than females). Hence, there was *no* time when sexual propagation did not include forced copulation (i.e., sexual assault).

Because humans and other hominins are in the minority of sexually dimorphic animals, the role of forced copulation in hominin evolution deserves special attention. The wildly disparate impact of hominins on their environments, the political structures that enable their impact, and the views of monogamy that emerge, allowing them to coordinate their impact, basically outlines what most would call a history of human sex. The motivations for sexual behaviors quickly move from mere survival into the political realm: how families, clans, tribes, and nations structure their power and its succession over generations. Forced copulation therefore creates a political difficulty in hominin evolution if rape underwrites the history of humanity—its nations, laws, languages, and governments—as we understand it.

Third, the current metaphysical orientation of the cosmos is similar if not identical to the physical laws throughout the history of the universe and will persist indefinitely. Scientists (Christian and not) and theologians may hold different views on the metaphysical descriptions that explain our current physics. For instance, Christians who are scientists might offer different answers to the following questions from their atheist or deist counterpart:

1. Does a divine power hold together all of the mass in the universe?

2. Is point gene mutation metaphysically random or divinely guided?

3. Was the overall course of hominin development aimed at one goal: humans?

4. Was there a historical garden at Eden in which God made humanity discontinuous with hominin development outside the garden?

Jewish or Christian scientists and theologians might not necessarily answer these differently from their atheist or deist colleagues. And each question could be debated between Christians of every tradition. I want only to point out here that the metaphysical answer will have theological implications for the physics and physiologies of our cosmos.

If we believe the current metaphysical arrangement of our universe is identical and unchanged throughout history and prehistory, then this belief alone might not be reconcilable to the conceptual world of the biblical authors.

Concepts and paradigms from the intellectual world of Scripture. First, there could have been at least one time and place—the garden in Eden— where scarcity was not. There was a time and place where competition-fueled violence was not a necessary reaction to scarcity. There was a time and place when humans were biologically well-suited to creation: to flora, to fauna, each other, and to God. It was the metaphysical arrangement of the created order that made it so (i.e., structure and direction were rightly oriented to each other).

Second, there was a time and place when sexual generation was properly instituted as singularly paired humans, in cooperation, and for the purpose of filling, tending, and producing children in league with the environment into which they fit. All later biblical texts on sexuality and marriage depend on this primeval sexual arrangement in Genesis 1–2. Within Scripture's logic, this was also a time when the act of reproduction and generation was wholly conducive to the human body. Conversely, there was a time and place where forced copulation (i.e., sexual assault) was not.

Third, the metaphysical orientation of the cosmos is presently dissimilar but not discontinuous with the physical laws throughout the history of the universe. The corrupted metaphysical orientation of creation will eventually be reoriented to beneficent effect for those fit to enter the new heavens and earth.

Fourth, various biblical authors *may or may not* have believed that (1) there were other hominins prior to the time of Eden, (2) there were other *Homo sapiens* outside Eden that were not human (e.g., neither *Homo sapiens* nor *Homo sapiens neanderthalensis*), (3) there were animals and plants outside the garden at Eden, and (4) there was a history of biological life, death, predation, and pain, prior to the garden. On these topics, the biblical authors remain largely silent.

A MORE PROFOUND CONFLICT

The difference between the Hebraic conception of the universe and commonplace evolutionary models should now be more obvious, even if complicated. It is not that they merely have competing histories, one natural and one religious. More problematically, they conceive of two different metaphysical and political schemes: one free of ethical considerations until they later emerge through evolution; the other one intertwining metaphysics with ethics from the start (see Gen 1:4-31; 3:16-21; 4:7, 11-12). For Hebraic thought, how humans behave in the world metaphysically reconfigures the cosmos.

While various versions of theistic evolution walk a different path from the commonplace evolutionary story, it is difficult to find a way in which they will not all ultimately conflict with the biblical authors' intellectual world. And this intellectual world is not referring to the model of a three-tier cosmos of the deeps, the land, and the heavens. It refers to how they conceptualize reality, how they analogically reason, and how they foster abstract thought.

I have explored the unique ways in which the biblical authors argue about the nature of reality itself. They show themselves to be keenly aware of the same themes in creation that Darwin will later notice and explore. The curse of scarcity implies violence as an inevitable response. Yet, the biblical authors focus on the ability to distinguish violence as a nonnecessary outcome of scarcity (e.g., Cain, Noahic laws, Mosaic laws, etc.). This seems to be the closest point of intellectual overlap between some cooperative views of natural selection and the biblical texts.

Genesis places sexually differentiated humans at the fore of creation and roots them in the image of God. The biblical authors restrain human sexual practices to only those that are cooperative and monogamous, an inefficient and ineffective way of propagating in evolutionary terms. Moreover, placing God as the supervisor of evolutionary processes does not resolve the problem I am highlighting here.

Some theistic evolutionary models ask us to scientifically sidestep some of the content of these creation texts by categorizing them as analogical, functional, or mythological explanations. As Dennis Venema and Scot McKnight say, "I'm not sure we are being respectful or honest with the Bible . . . when we make Genesis 1-2 speak about something it was not speaking about—biology, transmission of sin, genetics, and the like."[32]

---

[32]Dennis R. Venema and Scot McKnight, *Adam and the Genome: Reading Scripture After Genetic Science* (Grand Rapids, MI: Brazos, 2017), 108.

Certainly, the creation texts in Genesis are at times analogical, functional, and mythological (none of which requires them to be purely fictional). No doubt, Genesis's creation accounts were aimed at some other purpose than pure history and biology in our sense of the terms. We may agree on this description of Genesis 1–11, but that does not resolve the problem I have dredged up here.

I have shown that the biblical authors are arguing beyond a literalistic history of creation, even if they intended Genesis 1–11 to be read literalistically. The biblical texts demand that we enter their intellectual world and allow them to explain to us the nature of reality, not merely the physics or history of it. Discarding both the historical value of the text and the philosophical value of their arguments about the nature of reality seems to ignore everything needed to make sense of Israel, the resurrection of Jesus, and the empire he brings.

Even if we discard the historical-dramatic aspect of the texts—that history somehow unfurled movie-like as described in the texts—we are still left with a more profound conflict between the biblical authors and most evolutionary accounts, theistic or otherwise. The biblical texts conceptually enmesh the political metaphor of God as king over the cosmos *with* the moral obligations of humans and animals *with* the metaphysics of a blessed (natural) or cursed (disoriented) physical world from the beginning. Any Christian theological account of evolution must reconcile this sophisticated Hebraic entanglement of politics, morality, and metaphysics with its evolutionary model. Joshua Swamidass believes that his model—Eden as a divinely curated island-of-a-garden at the tail end of an evolutionary production of humans—meets these criteria.[33] I have given reasons from within the biblical authors' intellectual world to suspect that Swamidass's proposal still has some conceptual hurdles to overcome.

In the end, the question posed by the biblical authors is: Can the cosmos be metaphysically sublime or later renewed from ruin without God as king? God creates and sustains the food, the conditions to bear it, the earth to receive seed (and not receive human blood, Gen 4:10), and the celestially marked seasons to renew it. The very possibility of the eschaton

---

[33]S. Joshua Swamidass, *The Genealogical Adam and Eve: The Surprising Science of Universal Ancestry* (Downers Grove, IL: IVP Academic, 2019).

described by Ezekiel, Isaiah, Daniel, Malachi, Jesus, Peter, Paul, and John relies on a particular divine political order tied directly to the moral conduct of humanity and the metaphysical arrangement of our world. This political-moral-metaphysical arrangement began in creation, and it is pursued throughout biblical history to its natural ends: the renewed heavens and earth.

Putting aside the historicity of the creation—the what-actually-happened question—significant and irreconcilable tensions remain between the Hebraic conceptual world and that of some evolutionary sciences.

NOW WHAT?

I have argued that the intellectual world of the biblical authors makes our world existentially, ethically, and physically coherent.

For those who privilege the intellectual world of the Bible as guiding and generative for our own conceptual world, where do we go from here? The biblical authors swam in regional philosophical and literary currents of the ancient Near East filled with differing ideas about reality. They wrote forcefully about what they thought while eyeballing an array of metaphysical schemes and ideas about naturalism that surrounded them. Those peer traditions diverged widely from each other and the Hebraic view. This quandary about the exact narrative of nature or human origins is not new to them. They choose to take up neither the kind of intellectual exploration that their peer traditions expected nor one that our culture expects today.

Much like the details surrounding the political structure of Israel, the polity of the church, instructions for getting married, how to practice Sabbath, how to perform sacrifices, the purpose of stars, and more, details and procedures are left carefully unspoken across Scripture. Unspoken does not mean that they were silent on these matters. My goal has been to discern, from broader structures of thought, how the Hebraic intellectual world would have construed the inner workings of these social, physical, and metaphysical affairs.

I have looked for the drumbeat of thinking on the central matters of Darwin's theorizing. The goal is not to merely fill in the gaps of biblical reasoning with our own stories (as many Jewish and Christian interpreters have done over the millennia) but rather

- to discern what seems obvious, nodal, and guiding in their thinking on parallel topics today

- to notice that the Hebrews in the Old and New Testaments—alone among their ancient Near Eastern and Greco-Roman peers—cared to explain *as much as Darwin did* the connections of scarcity, fit, and propagation

- to understand that ancient Hebrews theorized about origins beyond the creation account into their national history and ultimately into the history of the cosmos

- to listen and learn from them in order to think more carefully about the proposals from the evolutionary sciences today

As for me, I retain a healthy and hopeful agnostic curiosity about the integration of these two conceptual worlds—evolutionary and Hebraic—whether they can be reconciled and how so. I hope that I have given a meaty enough account for colleagues to join in, correct me, or reconcile these potential areas of conflict as I have outlined them here.

SCRIPTURE INDEX